New Worlds,
New Technologies,
New Issues

Research in Technology Studies

Series Co-Editors

STEPHEN H. CUTCLIFFE
STEVEN L. GOLDMAN

Editorial Advisory Board

ALEXIS ALDRIDGE
EASTERN MICHIGAN UNIVERSITY

PAUL T. DURBIN
UNIVERSITY OF DELAWARE

MELVIN KRANZBERG
GEORGIA INSTITUTE OF TECHNOLOGY

EDWIN T. LAYTON, JR.
UNIVERSITY OF MINNESOTA

DONALD MACKENZIE
UNIVERSITY OF EDINBURGH

CARL MITCHAM
PENNSYLVANIA STATE UNIVERSITY

JEROME R. RAVETZ
UNIVERSITY OF LEEDS

JOAN ROTHSCHILD
UNIVERSITY OF LOWELL

LANCE SCHACHTERLE
WORCESTER POLYTECHNIC INSTITUTE

New Worlds, New Technologies, New Issues

Research in Technology Studies,
Volume 6

EDITED BY

Stephen H. Cutcliffe, Steven L. Goldman,
Manuel Medina, and José Sanmartín

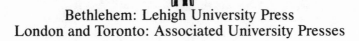

Bethlehem: Lehigh University Press
London and Toronto: Associated University Presses

Associated University Presses
440 Forsgate Drive
Cranbury, NJ 08512

Associated University Presses
25 Sicilian Avenue
London WC1A 2QH, England

Associated University Presses
P.O. Box 39, Clarkson Pstl. Stn.
Mississauga, Ontario,
L5J 3X9 Canada

The paper used in this publication meets the requirements
of the American National Standard for Permanence of Paper
for Printed Library Materials Z39.48-1984.

Library of Congress Cataloging-in-Publication Data

New worlds, new technologies, new issues / edited by Stephen H.
Cutcliffe . . . [et al.].
   p. cm. — (Research in technology studies ; v. 6)
   ISBN 0-934223-24-6 (alk. paper)
   1. Technology—Social aspects. 2. Technology and state.
I. Cutcliffe, Stephen H. II. Series.
T14.5.N465 1992
303.48′3—dc20   91-58102
                                  CIP

Contents

Foreword

STEPHEN H. CUTCLIFFE
STEVEN L. GOLDMAN

This volume is the sixth in the series *Research in Technology Studies* published by the Lehigh University Press. Volumes in the series are intended to present current scholarship addressing issues related to the interactions between science, technology, and society. Each volume in the series focuses on a topical theme set by the guest editor, or in this case editors.

The essays collected here were drawn from a Hispanic-American conference "Nuevos Mundos, Nuevas Tecnologias, Nuevas Perspectivas" organized by INVESCIT (Science & Technology Studies Center, Spain). The conference was held in the city of Valencia in December of 1989, with support from the Commission for the 500th Anniversary of the Discovery of America (Regional Government of Valencia). The objective of the conference was to provide a public forum for discussing social problems associated with the implementation of new technologies, including the cultural implications of technology transfer between societies. Consistent with INVESCIT's mandate, the conference participants included politicians, civil servants, and public interest activists, as well as academics and academic administrators, from Spain and from North, Central, and South America.

INVESCIT is a consortial research center for the interdisciplinary study of the relationships between science, technology, and society (STS). It was founded in 1985 and is supported by, and draws its researchers from, the Universities of Valencia, Barcelona, Basque Country, Balearic Islands, and Oviedo. Through two series of books, *Nueva Ciencia* (New science) and *Tecnología/Ciencia/Naturaleza/Sociedad* (Technology/science/nature/society), and special editions of the journal *Anthropos*, INVESCIT pursues a deliberate research program marked by the following platform: rejection of uncritical theoreticism and thus of the subordination of technology to science; support for a "deep ecology" stance vis-à-vis the environmental risks posed by all technologies, new or in place; the addition of social impact assessments to risk-benefit calculations; the necessity of a historical analysis of the relationships between theory and practice; a critical analysis of technological practice and of the presuppositions underlying the mathematical apparatus, and the putatively value-neutral (because objective and quantitative) character of modern science.

The essays in this volume are grouped around the three themes announced in the conference title: the crystallization of new (cultural) worlds around new technologies, the challenges to democracy posed by the form of the implementation of new technologies, and the precipitation by modern technologies of new social and moral issues. Each of the essays is either a revised version of a paper invited for presentation at the conference, or was subsequently invited from a conference participant. Each theme is introduced by a condensed version of the comments offered by the moderator of the conference sessions devoted to that theme. The volume as a whole reflects the international concern with the integral role played by modern technologies in a rapidly changing world.

Introduction

STEVEN L. GOLDMAN

A characteristic of all of the presentations at the conference New Worlds, New Technologies, New Issues was their unqualified, and unapologetic, confidence in the causal significance of the social relations of technology. There is no arguing the social relatedness of technology here. That is taken as established. That technological innovation is a social process serves these authors as a point of departure for attempting to discover how that process is determined by the social structure of technological action and not from technical knowledge conceived as possessing an extrasocietal nature of its own that dictates its social impact. These essays thus reflect a kind of "second generation" consciousness on the part of STS scholars. They also point to an emerging "third generation" consciousness, in which technologies themselves are social creations, at least as much expressions of the values and institutional structures characteristic of the society in which they appear as they are expressions of the properties and "laws" of natural phenomena.

NEW WORLDS

George Bugliarello, alone among the authors addressing the theme "New Worlds," sees in new technologies crowding in upon us a bright promise for a new and better world. Albert Borgmann, Manuel Medina, and José Sanmartín are unsettled by the "new" world already created by modern technology, that is to say, our world, the world of the so-called advanced societies, which today suffers from a host of moral, political, and physical ills caused by science and technology. They also express profound suspicion of the future of that world, unless the cultural foundations of technoscience are undone and new foundations are put in place that will allow beneficial relationships to flourish among ideas, machines, and values.

9

Bugliarello, by contrast, distinguishes science and technology as knowledge bases from the social contexts of their application. The latter, lacking critical frameworks adequate to orienting societies to science and technology constructively, are the causes of the problems ascribed to technoscience. Thanks to science and technology there *has* been progress since 1492, and the promise of emerging technologies such as "hyperintelligence"—the synergistic linking of individual human, social, and artificial intelligence—is for greater progress still. But to realize that progress we need to cultivate an optimistic attitude toward modern technology, acknowledging its intrinsic benevolence and dismissing false ideas, such as the claims that technology enslaves us, that there are two cultures with the mass of humanity on the outside of science and technology, and that the unconditional elimination of modern armaments would bring peace to the world.

In fact, Bugliarello proclaims, modern technology has been more often liberatory than not, the world is being drawn into a single "technological" culture, and mass destruction weaponry has been a cause of the lack of major wars since 1945. Granted, new technologies are precipitating new worlds, and in those worlds technologies may be abused to the detriment of the masses and of nature. But this only isolates a challenge to philosophy: how to create a framework for the "intelligent development" of new scientific knowledge and new technologies. Because technology, for Bugliarello, is the "extracorporeal product" of human evolution, it is coordinate with, and not in opposition to, society. What the philosophy of technology needs to do is to devise "fundamental tests to which technology must be put to determine its validity," that is, to determine the "correct" proportion, the "harmonious" balance, between people, social institutions, and machines.

This may sound like the calls for social reconstruction in the essays by Borgmann, Medina, and Sanmartín, but it is not. For them, as for Ihde, the spheres of technoscience and of society are mutually penetrating, mutually determining, and inseparable in fact, though not in thought. The reconstruction they call for is thus a total re-construction, a re-construal, of society, of technology, and of their interrelationship. This is a call to radical, if not to revolutionary, action on social values and institutions and, *eo ipso,* on science and technology, which ineluctably will be transformed in the process. For Bugliarello, the sphere of technical knowledge is not itself determined by the structure of the sphere of social action. It is what it is; only the transfer of technical knowledge to the social sphere is a function of social values that determine the form of its implementation. Thus, for Bugliarello, an intellectual challenge to philosophers replaces the revolutionary reconstruction of society that Borgmann, Medina, and Sanmartín argue is the only path of hope in the face of the problematicity of contemporary technoscience.

For Borgmann, modernity, the realization of ideas promoted by Bacon, Descartes, and Locke, is already undergoing a transformation, but the form of postmodernism is so far ambiguous. Postmodern society may become "hypermodern," a society in which the already pernicious influence of modern technology will become still more pervasive and dominating. Or it may be "passing through" domination by modern technology to a "metamodern" state, a state in which technological action will be "context sensitive and historically reverent," carefully and respectfully (rather than aggressively) realistic, attentive to different "voices of reality."

The "modern project" entailed the "fusion of the domination of nature, primacy of method, and the sovereignty of the individual." Today, postmodernism already rejects Baconian realism, Cartesian methodological universalism, and Lockean liberal individualism. Borgmann finds hints in postmodern architecture and economics of the values he associates with "metamodernism," and that is hopeful. But the hold of modern technology on social institutions and on values that were created to abet its domination is strong, and there are hints of emerging "hypermodern" values, too. Information processing technologies, for example, have equivocal social consequences. They can oppose the pressures of modernity, or they can be used to create a comuter generated "dematerialized unreality" that stimulates a "feverish pointlessness and restlessness" among their widening circle of users. In short, Borgmann projects modern society into a kind of activist Heidegerrian drama in which the possibility exists that, by choosing well, we can change the form of our being in the world. The choice, appropriately in a Hegelian way, is created for us by tensions internal to modernity. In this we are passive. But we can play an active role (*contra* Heiedegger) in determining which of the choices will be realized.

Sanmartín and Medina share Borgmann's judgment that modern "technoscience" is the root of many of the more threatening ills of contemporary society. Technoscience is pathological, for Medina, because of its "theoretistic" bias, the privileged status attributed to theory as superior to practice, standing outside history, culture, and values. This status insulates science from responsibility for the social and environmental consequences of its applications. It also precludes a correct understanding of technology, so we are unable to respond effectively to the "culture of risk" that technoscience has created. We are pressured to respond to technology related environmental, social, economic, and political problems by turning to science for solutions. But this is precisely the one avenue guaranteed not to solve those problems, because it is itself implicated in causing them!

To respond "decisively" to these problems, "it is essential that the technoscientific cosmovision itself be replaced." This is above all a political enterprise, entailing the creation of a new world by way of a "radical renovation of the spheres of technological action, social organization, and cultural values." Central to accomplishing this is replacing the objectivistic, value-neutral, theoretical understanding of nature underlying technoscience with a new, "operational" understanding of nature. Echoing Borgmann's notion of "metamodernism," this operational understanding will be "non-theoretical and praxis based . . . firmly rooted in historicity, in concrete social and environmental contexts, and in the complex factors motivating human activity." The upshot will be, in effect, an inversion of the traditional conception of the relationship of science to technology. On the "operational" view, practice is primary, and theory derives from practice; technology, value laden, historically situated, interdisciplinary, physically and socially context specific, becomes primary and science secondary.

Sanmartín agrees with Medina (and both with Dewey) that science is a species of technology, and with Borgmann that the most pressing problems confronting society are the result of "enlightened modernity . . . a civilization whose superideology had science as its backbone." Modern science is a cognitive activity, but also a source of power, especially the power to perfect and control the "enemy," Nature. Modern "industrial" technology has a coordinate superideology, associated with the ideas of progress, of the imperative of technological innovation, of technological determinism, and of military technologies as the basis of physical and political security. Anchored in this superideology, modern technology "recklessly" exploits natural resources. And it concentrates industrial production and economic power in ways that generate unsustainable pollution levels and that are socially destructive because they promote standardization of personal and social behaviors.

Sanmartín, like Borgmann, sees some signs in emerging postmodernism of opposition to this superideology, especially trends toward deconcentration and destandardization, but unless the superideology itself is abandoned, these trends will be overwhelmed. Tomorrow will either be a "hypertrophied version" of today, or a new view of technology will have to become pervasive. In that case, a new world will be created and if, on this new view, technology is identified as a means of achieving socially established goals that are environmentally sensitive, decentralized, and, to borrow a phrase of Don Ihde's, "pluricultural," then the new world will be a better one than ours. To promote the dissemination of this new view of technology, Sanmartín believes it is essential that the imperative character of technological innovation, and the putative autonomy of science and technology, be attacked at every turn. This is the destructive phase of the transition from the old to a new superideology. The constructive phase is the need for embedding scientific and technological practice in a truly informed public awareness of the natures of those practices.

Ihde agrees with all of the other authors that the contemporary "lifeworld" is "highly technologically textured." "Euro-American technoculture" has by now affected every culture in the world, but Ihde is less interested in judging the quality of life in that technoculture than in understanding how technologies transform lifeworlds. To that end, he reflects on three historical voyages: Columbus's "daring, false-belief-situated" voyages, the voyages of the Polynesians who settled the far-flung Pacific Islands, and the voyages of the Chinese admiral Cheng Ho around the eastern and southern Asiatic coasts to the Persian Gulf. Why, he asks, was only Columbus's voyage a watershed, on "our" side of which a new world opened up characterized significantly by a "science embodied in material technology?"

It is tempting to dismiss the Polynesian voyages as inconsequential because of their "low technology" character: no maps, no compass, indeed, no "instruments" at all, no mathematical navigation techniques. But Ihde argues that this is a highly ethnocentric reading of technology. The Pacific Islanders in fact employed highly sophisticated, albeit nonartifactual, navigation "technologies," which Ihde describes, while their ships were in certain respects superior to those of contemporary Europeans. Nevertheless, these culture-disseminating and population-dispersing voyages were of no great consequence globally, and of little enough consequence even locally.

Cheng Ho, on the other hand, wielded a set of technologies manifestly superior to that of Columbus. His fleet and its crew dwarfed Columbus's, as his nine-masted flagship, a treasure ship, at 444 feet, dwarfed Columbus's largest ship, the Niña, at 96 feet. Yet Cheng Ho's voyages, too, had no lasting impact; they precipitated the emergence of no new world, while Columbus's did. Ihde concludes that "technologies in the ensemble are distinctly embedded into cultures," and their impact on societies cannot be understood in isolation from that embedding. What Columbus carried with him was not a decisively superior collection of material technologies, but a set of values and attitudes towards technologies and their uses. In particular, Columbus carried an "overhead, or god's-eye, view" of the world that reflects Western cultural values of dominance and control.

Unlike Bugliarello, Ihde rejects the notion that there is a technosphere independent of the sphere of culture. With Medina, Sanmartín, and Borgmann, Ihde locates the dynamics of technological action in the value structure of its social embedding. But like Bugliarello, and unlike Medina, Sanmartín, and Borgmann, Ihde does not call for disowning modern technology and beginning the reconstruction of the social sphere in order to create a new technology and through that a new and better world. Ihde is closer to Heidegger than Borgmann in recognizing the emergence, unwittingly, of a new vision of being that could lead to a new world. As Euro-American technoculture has spread to other cultures, the points of contact have caused a reflux, a flow of transformative cultural influences back onto Euro-American culture. The result is that today ours "is a pluricultural vision" of the world that is latently deeply syncretistic and thus post-modern in its direction. It is a "compound way of seeing," a "bricolage of pluriculturalism" that contains the seed of a new world and thus a new embedding for future technological action, and the potentially culturally transformative reactions those actions may cause.

NEW TECHNOLOGIES

Industrial technology has "transformed—revolutionized—human institutions, thoughts, and actions," in the main beneficially, in Melvin Kranzberg's opinion. At the same time industrial technology is directly or indirectly responsible for serious current problems. How, Kranzberg asks in his introductory remarks, are we to respond to those problems without abandoning modern technology, which is linked to whatever personal and social progress there has been in the modern era? His reply is that while we have no hope of solving those problems without technology, more technology will not by itself solve these problems either, for they involve "human values, social organizations, environmental concerns, economic resources, political decisions, and a host of other socio-cultural factors." The root problem confronting modern societies, and still more acutely societies striving to become modern, is not ignorance of relevant technical knowledge. It is the far more difficult problem of how to factor complex, competing, and conflicting social considerations into public policy decisions affecting the directions of scientific research and technological innovation. The essays in this section of the book revolve around just that theme: political responses to new technologies.

Paul Durbin, for example, begins with profound pessimism, but ends on a note of qualified optimism. The fulcrum of Durbin's essay is Daniel Bell's monograph *The Cultural Contradictions of Capitalism,* and Durbin lays out his own thesis against the backdrop of Bell, quoting Nietzsche on nihilism as the inevitable end to which technology is driving modern Western culture. To this gloomy assessment, Durbin adds Marcuse's final despair of public activism as capable of ameliorating the one-dimensionality of modern culture, and Ellul's conclusion that only religion can save modern societies from the totalitarian imperialism of technique. To this negativism, Durbin responds that we can "create new, meaningful symbols, a new culture for our troubled technological world," but only if technical experts "work alongside other activists trying to make ours a better world . . . in which as many human beings as possible find a fuller meaning for their lives." The struggle to accomplish this must, for Durbin, be embedded in a Deweyan program of action if it is to be successful.

Durbin surveys three recent interpretations of Dewey's philosophy that point to its fertility for guiding the creation of a new culture, one in which technology's impact will be more beneficient than at present. All three dismiss instrumentalism as the correct reading of Dewey's analysis of technology. Thomas Alexander makes aesthetics the central motif of Dewey's pragmatism, emphasizing the communalistic and contextual character of the guidance Dewey provides for progressive social activism. Larry Hickman argues that an "explicitly and consciously . . . meliorist critique of technological culture" is the key to Dewey's philosophy. Cornel West interprets Dewey as providing the foundation for a "prophetic pragmatism" that can illuminate the historical consciousness of particular communities struggling to find "new cultural meanings against entrenched traditions."

"Are there," Durbin asks, "activists around in sufficient numbers to carry on the struggle that West calls for?" As did Borgmann, Ihde, and Sanmartín, Durbin sees modern culture in crisis, transforming into a new but as yet still ambiguous form. It is possible that technological nihilism will triumph, as Borgmann thought that postmodernism might become hypermodernism, as Ihde thought that pluriculturalism might be superficial, and as Sanmartín thought that a hypertrophied version of modernity might be emerging. But it is also possible to share, albeit cautiously, Bugliarello's and Kranzberg's optimism. Public activism, informed by a Deweyan pragmatic philosophical action program, may well be capable of creating a "new progressive, more open, technological democracy aimed at including the poor and the oppressed." If the outcome is uncertain, that is hardly surprising. After all, a "meaningful existence is not something to be taken for granted, but . . . must be won through arduous social struggle."

This notion of a struggle dominates Margarita Peña's account of the "development debate" in Third World and especially in Latin American nations. The debate is between promoters of economic development based on First World industrial technologies and the *dependentistas*, who support development keyed to indigenous science and technology. The *dependentista* position is that First World industrialism, and the technical knowledge coordinate with it, are ideologies that legitimate certain realities and certain political policies only, precluding local definition of social development. For the *dependentistas*, development keyed to modern industrialism entails allowing the quality of life in one's society to be determined by the fluctuating requirements of the First World dominated globalized industrial economy. Dependence, they recognize, is not necessarily backwardness. A nation like Brazil may develop a very sophisticated scientific and engineering capability, may even translate that into a globally attractive industrial capability, and still not solve its national development problems. If science and engineering education and research and development programs are dictated by First World interests, including First World professional recognition criteria, then a nation will remain dependent (on industrial technoculture), even if it can play the industrial "game" on even terms with industrial leaders. South Korea is an illustration of a nation that is in just such a position.

The *dependentistas* do not want to abandon the goals of progress and development, though that is, understandably, precisely what they are attacked as doing. They support, rather, the creation of locally generated technical knowledge "geared to the solution of the basic needs of the [local] population, such as housing, clothing, food, and health." This says nothing about the level of technology to be deployed in satisfying these needs. The *dependentistas* are not opposed to advanced technology as a matter of principle. Some of the latest microelectronic communication and information processing technologies, also biotechnologies, seem to hold exceptional promise precisely for Third World societies. But as a matter of fact, very little is known about the impact on premodern societies of suddenly incorporating such technologies. Nor is there any reason the First World creators of those technologies should be studying those impacts. That is properly the business of the premodern societies themselves that are, on the contrary, being pushed by many promoters of development into adopting industrial technologies as if development and progress were automatic concomitants of the presence of those technologies in a society.

The alternative is the integration of locally generated technical knowledge into social and economic policies that reflect locally determined development agendas. Peña asks whether a Latin American-style technology is possible: "That is, can we have all the advantages of development without having to give up the customs, cultural values, and ways of life so profoundly rooted in [Latin American] tradition and associated with a quality of life that has already been lost . . . in the most advanced countries?" The examples of the maquiladoras of northern Mexico and the industrial "free zones" in Asia are not encouraging. What is lacking throughout the Third World is a critique of technologies as forms of life, in the manner of Ellul and Winner, rather than as tools to be used but otherwise ignored as inconsequential.

Worthington's essay is a perfect complement to Peña's, focusing as it does on the globalization of production as the current stage of the most advanced industrial economies. Worthington examines from within First World societies the same phenomenon that Peña responds to from the perspective of Third World societies that are the object of this latest form of imperialism. At the same time, Worthington recalls Durbin's essay by concluding that the adoption of a "pragmatic holism" is the key to undoing the reductionism that is the epistemological ground of the modern industrial nation-state.

Modernism is identified by Worthington (as by Borgmann) with sundering the connectedness people historically felt with society, nature, and the cosmos in favor of rampant individualism. The result has been a progressive alienation of people from one another and from nature, and a "compulsive resort" to ever more powerful technologies in order to resolve problems created by this alienation. The resolution lies in the reintroduction of holism, and while holistic critics of modernism have appeared, so far they have had little impact. Ironically, the most dynamic source of holistic values today is advanced industrialism, which is trying to create a globally integrated industrial system.

The irony is that, for Worthington, the growth-driven political economy of advanced capitalism is a fundamental obstacle to a holistic response to the alienation-linked social concerns caused by modernism. The feminist, peace, and ecology agendas, for example, can only be accomplished by transforming the forces of production and authority relationships in modern societies along holistic lines, and the prevailing vested interests are hardly likely to cooperate in this. At the same time, however, these same interests are promoting holism through their self-serving industrial globalization projects. But the globalization of production is not equivalent to globalization of progress. The "channels" of industrial globalization are dictated by value judgements deriving from "old means of understanding the world," that is, from an understanding of the world anchored in the modernism that created the First World–Third World disparities as well as the systemic problems that today afflict First World societies themselves. Furthermore, globalist rhetoric suffers from a fundamental confusion between the truly global (and thus holistic) and the merely international, which remains a vehicle for parochial values and policies. This confusion must be clarified before we can constructively "address the ethical and political challenges of the world we have created."

Goldman's essay also deals with a systemic feature of modern technology, and also from within "developed" societies. He argues that the process of technological innovation is driven not by "objective" technical knowledge, but by subjective managerial decision making. If we assume that in a democratic society the public is entitled to some effective participation in affairs that affect the conduct of their lives, then they are entitled to participate in creating technological innovation policies. But if innovation is driven by managerial decision making, then only through involvement in this process can the public play a significant role. Reacting to innovation decisions after the fact, through regulatory agencies or recourse to litigation, is largely ineffective and always inefficient.

The problem with public involvement with managerial decision making is that the latter is protected as proprietary, especially in an entrepeneurial economy such as that of the United States. The more common argument that the public cannot play a role in technology policy or innovation decisions because they lack the requisite expertise, is belied by the fact that the overwhelming majority of those who make those decisions also lack technical expertise. The upshot is that unless the public is to be locked out of that aspect of modern society that is most responsible for social change, namely, technological innovation, we need to create institutional mechanisms adequate to the realities of the innovation process. Like Peña and Worthington, Goldman sees this as implicating at a fundamental level the existing relations of power and authority in modern society. If this is an intimidating challenge for critics of innovation, any response to the social impact of new technologies that leaves these relations undisturbed will almost certainly be merely palliative.

NEW ISSUES

Each of the essays in "New Worlds" and "New Technologies" argues that changes in values and institutions are preconditions of a more beneficient relationship among people, nature, and technology. But all of the references to values and institutions are extremely broad: superideologies, cosmovisions, metamodernism, pluriculturalism, pragmatic holism. The essays in "New Issues" are more narrowly focused: on the ethics of military technologies, on the values of radical environmentalism, on public literacy in science and technology, and on democratic legislative mechanisms. The specificity of the value commitments in Elena Lugo's introduction signals this shift in focus from the general to the particular.

Self-liberation "by way of scientific-technical knowledge is now possible" in principle for all citizens of the modern world, but it is at the same time impossible for almost everyone because the scientific-technical conception of the world lacks all "foundation in the order of meaning, of being, and of values." As the explanatory power of modern science grows, and as the power of modern technology over nature increases, it becomes necessary for society to assume explicit responsibility for the conduct of science and technology, and for their consequences. But the very nature of the world that modern science and technology have created entails that this responsibility falls on a "human collectivity that unfortunately lacks certainty, integrity, and solidarity regarding a common good, the meaning of personhood, and the value of community."

Lugo's vision of a properly responsible posture toward technology requires that we pursue objectivity in knowledge, that we use technology to preserve "inner freedom" in the first instance and to effect the utilization of things only second, and that we remain open to a transcendental order in being. In this way, technology can ground a "new wisdom beyond mere rules of prudential calculus of risk and benefit." By "objectivity in knowledge," Lugo does not mean merely the impersonal, that which is independent of the individual knower. She refers, instead, to a dialectical interconnection of all things in nature, including human beings, an order in which the power of nature over us is "counterbalanced by a spiritual penetration" of nature by us. By acknowledging our integral participation in such an order, Lugo believes, we can find an "objective" basis for renouncing our unlimited power over and domination of nature.

To effect such a responsible attitude requires that technology be a public concern, and the Quintanilla and Ten essays address the political and educational implications of this requirement. But Lugo observes that a generic implication is the need to resist the depoliticization of the masses that seems to accompany the current scientization and technologization of politics. Without the participation of the masses, there can be no responsible social posture toward technology. Furthermore, this responsibility entails balancing the intensely masculine values that have informed the development of modern science and technology with values expressive of a "feminine principle" that "can, and indeed must, exist in every human." Lugo looks especially to the work of Carol Gilligan and Sara Ruddick for the articulation of these values, which include: a heightened moral sensibility, overcoming alienation from our own bodies, recognizing the priority of holding over acquiring, resistance to the seductiveness of idealistic abstractions, rootedness in the concrete and the empathic.

The widespread environmental consciousness that began to emerge in the late 1960s is one expression of a persistent public concern with technology. In an analysis that echoes themes in the Worthington, Medina, and Sanmartín essays, Michael Zimmerman examines the political values underlying the deep ecology movement founded by Arne Naess. Zimmerman is particularly concerned about attacks on deep ecology as "ecofascism," about charges that deep ecology is racist and authoritarian, about feminist criticism of the movement, and its relationship to Heidegger's philosophy.

From its inception, deep ecology has distanced itself from "mainstream," reformist, enivronmentalism, for the same sorts of reasons that Karl Marx distanced himself from Fabian socialism. The modern industrial system cannot fail to destroy the ecosphere. That is the fundamental conviction of deep ecologists; all of their other commitments pivot about that "fact." To work "within the self-destructive logic of the technological system," as the reform environmentalists do, is to be co-opted by that system. What is necessary is "nothing less than a complete transformation of human understanding . . . if humanity and the ecosphere are to survive in the long run." This transformation must begin with "critical reflection on the metaphysical presuppositions" that have led the West, and through the West the entire world, to its current state.

The values responsible for jeopardizing the continued existence of the ecosphere within which humanity evolved include anthropocentrism, atomism, dualism, and a narrowly utilitarian attitude toward nature. Anthropocentrism, in turn, is attributed to the Graeco-Roman cultural tradition, especially Stoicism, to certain strains in Judaeo-Christian ethics, and to the humanist ideologies of both capitalism and Marxism. The new values that we need to internalize include ecocentrism, relationalism, and a holistic ontology from which a holistic ethics can grow.

Some social ecologists and feminist environmentalists have criticized the deep ecology focus on anthropocentrism as the central villain in the "eco-drama" currently being played out. They argue that hierarchical, patriarchal, and authoritarian values underlie the Western attitudes that have put the ecosphere in jeopardy and that these are the values that must be eliminated. Naess himself centered his program for transformation on a Buddhistic notion of Self-Realization—in which personal fulfillment entails identifying with the totality of being—that has strong Heideggerian overtones. Zimmerman, who recently published a detailed study of the value structure of Heidegger's philosophy, concentrates on this dimension of the deep ecology platform.

There are, Zimmerman argues, similarities between deep ecology's critique of Western metaphysical presuppositions and Heidegger's critique of Western productionist metaphysics. Heidegger's notion of *Gelassenheit,* for example, "leaving things be," is not unlike Naess' notion of Self-Realization. But Heidegger rejected the idea of locating humanity in nature and expressed strong antinaturalist sentiments in his writing, both profoundly inconsistent with deep ecology commitments. In addition, Heidegger ignores the "authentic emancipatory dimension" of the Enlightenment—that all humans are worthy of respect and that cultural differences should be celebrated, not eliminated—for all of its limited technoscientific and political vision. And Heidegger's ties to National Socialism's political objectives and its ideology must be cause for concern about tendencies in the deep ecology movement. What makes Zimmerman optimistic about the movement, in spite of recent misanthropic and racist remarks by some members, is that the bulk of its writings are anti-authoritarian, that in spite of its hostility to modern science and technology as institutionalized, deep ecology (*contra* Heidegger) is open to science as in principle capable of illuminating being, and that deep ecologists in fact "adopt the liberatory language and hope of the Enlightenment" in spite of their criticism of it.

Carl Mitcham's concern is with the extent to which the socialization of modern technology implicates military applications, and a network of coordinate political values and national policies, domestic as well as foreign. Like Zimmerman, Mitcham addresses a radical minority critique of contemporary technology, in his case, the view that intensive technological development (of any kind) inevitably entails rampant militarism. The more common view, of course, is the reverse: rampant militarism pushes intensive technological development. On the latter view, the military-industrial (-academic) complex that has played so large a role in American, and world, affairs since the 1950s could be avoided by controlling militarism, while nonmilitarist technological development could proceed at any pace desired. On the former view, even the intensive development of new "reformed, humane, alternative, or ecological" technologies will evoke militarist values and policies.

Mitcham reviews the argument for the radical critique and agrees that untying the bonds between technological development of any sort and warfare may not be easy. It is not clear that science and technology can "have a reality independent of, and uncontaminated by, military imperatives and influences." He looks to Henri Bergson's *Two Sources of Morality* and Edmund Husserl's *Crisis of the European Sciences* for illumination.

Bergson and Husserl both argued, though on different grounds, that militarism deforms science and technology and both proposed correcting this deformation by an appeal to life forces. For Bergson, war is natural, but with the appearance of weapons of mass destructiveness the "elan vital" must assert itself and this will obviate warfare. For Husserl, the collateral development of militarism and modern technology reflects the inability of modern science to address the question of the meaning of human existence. The post-Galilean mathematization of nature created a moral vacuum that was filled by "might makes right," and thus the unbridled pursuit of might, to which the growing power of technical knowledge was bent. To undo this connection, Husserl urges a return to the *Lebenswelt* from which science and technology as well as militarism are absent. All three dissolve in the undifferentiated state of the primordial lifeworld. Mitcham offers a metaphor of the world as a stage on which institutional actors play out their roles in a science-technology-military drama whose current action threatens to destroy the very stage on which the drama is being played. The changes being called for by many in the world is respect for the continued existence of the stage, that is, of life. But it "remains to be determined whether the postmodern reaffirmation of life and delimitation of science, technology and the military is ultimately philosophic or nihilistic in character."

Lugo observed that public involvement in technology was a *sine qua non* of the resolution of current technology-generated problems, and Quintanilla's essay addresses this issue. Democratic institutions can only "maintain a legitimate capability" if there is broad public support for science and technology policies. To create such support, Quintanilla argues, modern societies must effect certain social institutional changes, and an "important role" for the media must be cultivated. Quintanilla surveys the challenges facing democratic political institutions, and reviews various approaches to technology assessment that have been emplaced or proposed. His commitment is to the efficacy of political will in spite of the problems posed by reform of legislative institutions. Thus, the legislature remains the "ideal place to make decisions regarding the social evaluation of technological development." A condition of making intelligent decisions, however, is that legislators need comprehensive, objective, information collected through an open, participative process, about relevant science and technology options.

Quintanilla is satisfied that current legislative experiments in a number of countries are moving in this direction, and that this movement is likely to continue. What is still lacking is the broad base of public support for political action that improved media coverage of these issues could crystallize. Mass media treatment continues to suffer from a stereotyping of science and technology that "devalues their potential social and political role." It needs, instead, to adopt a process interpretation of research and development, to depict science and technology as contingent social phenomena and thus essentially political, to stress the open nature of science and technology policy decisions, and to stress the legitimacy and importance of public participation in public participation in policy debates on these issues.

Antonio Ten focuses on a parallel route to public science and technology education, the creation of science and technology museums and centers. Ten begins with what he calls the "science-technology paradox," namely, the more science and technology used in a society, the less capable citizen action is in that society. The culture and the values of science and technology are today simply superimposed upon the traditional culture and values that dominate family life and the school systems in Western societies. Acceptance of this paradox contrasts starkly with the lip service paid in those societies to a universal right to "culture, information, and freedom, and the right to criticize." If, as Ten argues, the family and the schools cannot unite the culture of science and technology with traditional culture, then "alternative institutions for science education" are needed.

Ten offers a very brief, highly selective, history that reveals that the origins of the science museum lie in the earliest phase of the formulation of the mechanical philosophy of nature. In each century since the seventeenth, there has been a dramatic increase in facilities for the public dissemination of information about science and technology until recently, when a "veritable boom" has set in for "high tech" institutions whose aim is extending science to the general public. No standard model has emerged for such an institution, but Ten identifies three distinctive philosophies informing those that have been built so far: the scientific museum, involving scholarly collections, explicit learning experiences, active experimentation; the science center, filled with displays of scientific and technological phenomena that are explicitly entertaining and implicitly (it is hoped) educational; and frankly entertainment complexes.

The interaction between visitor and exhibit in these institutions is far more complex than in traditional museums, and there is a poor understanding today of what students, especially, take away from their visits and why. That these centers serve as an extension of the classroom is probably a poor model, and they are not a substitute for formal academic study. On the contrary, they seem best designed to be informal, open, relatively unstructured, and idea, rather than knowledge, oriented. Also, they are as much for teachers as they are for students. For the adult public, such museums should provide a source of information relevant to science and technology policy issues, a means of mass communication, and prods to reflection, on those issues.

"The world of science has always been an elitist and esoteric world," but that is no longer supportable given the nature of our science- and technology-related social problems. To create a wider, and informed, public involvement, a special effort needs to be made to formulate curricula for these centers that can communicate to naive visitors, using the latest technologies. And these centers must be brought to the people, somehow in spite of their size and complexity, perhaps by distributing satellite units around the periphery of major cities, perhaps by having itinerant centers. The crucial point is that these institutions must not be seen as cultural "frills," but as integral to modern political life.

An overview of the fifteen contributions to this volume reveals that two, those by Bugliarello and Kranzberg, are fundamentally optimistic about technology as a driving force behind personal and social progress. Both authors acknowledge that science and technology have become problematic for modern societies, but they believe that social adjustments can be made that will allow technical knowledge to flow into society less problematically. Ihde's essay adopts a nonjudgmental, seemingly technologically determinist posture locating the social impact of technology in the value-laden social contexts of technological action. He points to a new cultural "vision" that is already emerging as a consequence of a reflection back onto Western culture of non-Western cultural influences through the very same technological channels by means of which Western culture has influenced all other cultures in the world. This leads to the creation of a dialectical bonding among cultures that must eventuate in a true world culture of "compound vision," as Ihde calls it.

Three of the contributors, Goldman, Quintanilla, and Ten, argue the need for more or less serious structural changes in the socialization of science and technology. All are either well within the current system or seem not to stretch that system to the breaking point. Durbin's defense of public activism informed by Deweyan pragmatism entails a more radical change, as does Peña's defense of the anti-*dependentista* version of the form that institutionalization of modern technology must take in Third World nations, if their "development" is to be beneficial for *them*.

The other seven essays argue that only a revolutionary transformation of Western cultural values can prevent global catastrophe, precipitated by modern science and technology. All locate the destructiveness of science and technology in the values built into what Borgmann calls the "modern project," in the peculiar rationality, the "superideology" (Sanmartín) or "cosmovision" (Medina) of modernity, in the Enlightenment vision of human and social progress. Current, so called "postmodern," cultural trends are at best grounds for limited hope. These trends are ambiguous, at best pointing to the emerging possibility of a beneficient successor to modernism, but requiring such a profound transformation in prevailing values, and confronting the inevitable opposition of entrenched vested interests of enormous power, that these essays establish for the volume as a whole a mood of disquieting anxiety. And in this, too, the conference reflected a "second generation" STS consciousness, moving well away from the generality of many early STS studies and toward the formulation of a concrete political platform for informed social reconstruction.

Contributors

ALBERT BORGMANN is Professor of Philosophy at the University of Montana in Missoula. He has served on the board of directors for the Society for Philosophy and Technology. He is the author of *The Philosophy of Language* (1974), *Technology and the Character of Contemporary Life* (1984), and *Beyond Sullenness and Hyperactivity: Crossing the Postmodern Divide* (1992).

GEORGE BUGLIARELLO is President of Polytechnic University. He is an engineer and scientist with a broad background ranging from civil engineering to computer languages, biomedical engineering and fluid mechanics. He is founder and editor of *Technology in Society—An International Journal* and has authored some two hundred professional papers and several books.

PAUL T. DURBIN is a professor in the Philosophy Department and the Center for Science and Culture at the University of Delaware. He also teaches in the Ph.D. program on technology and society there, and at Jefferson Medical College in Philadelphia. He has edited *A Guide to the Culture of Science, Technology, and Medicine* (1980 and 1984), *Research in Philosophy and Technology* (8 volumes, 1978–85), and *Philosophy and Technology* (9 volumes, 1983–91), and has written a great deal on social responsibilities of scientists and engineers.

STEVEN L. GOLDMAN is Andrew W. Mellon Distinguished Professor in the Humanities at Lehigh University and a member of the history and philosophy departments. He is a former director of Lehigh's Science, Technology and Society Program. His research focuses on the social relations of science and technology in Western culture, and he is the editor of *Science, Technology and Social Progress* (1989), also published in the Research in Technology Studies Series by the Lehigh University Press.

DON IHDE is Leading Professor of Philosophy and former dean of humanities and fine arts at the State University of New York at

27

Stony Brook. His research focuses on phenomenological questions in the philosophy of technology. He is the author of numerous works including *Technics and Praxis* (1979), *Existential Technics* (1983), and *Technology and the Lifeworld* (1990).

MELVIN KRANZBERG is Professor Emeritus of the History of Technology at Georgia Institute of Technology. He was one of the founders of the Society for the History of Technology and edited its journal, *Technology and Culture,* for two decades. He is the author and editor of articles and books that pioneered the "contextual" approach to the history of technology.

ELENA LUGO is Professor of Philosophy at the University of Puerto Rico at Mayagüez. She directs the Center for the Philosophy and History of Science and Technology—an interdisciplinary unit of research and teaching of the UPR-M. Her most recent works are: *Medical Ethics* (1983), *Philosophy and Psychology* (1983), *Ethics for Engineers* (1985), and articles in the fields of biomedical ethics, ethical theory, and the phenomenlogy of Max Scheler.

MANUEL MEDINA is Professor of Science and Technology Studies at the University of Barcelona. He is a co-founder of INVESCIT (Science and Technology Studies Center). His research focuses on the multidisciplinary analysis of science and technology and their consequences for technology assessment and science policy. He is the author of *De la Techne a la Tecnología* (1985), editor of *Noves Tecnologies* (1986), and co-editor of *Ciencia, Tecnología y Sociedad* (1990).

CARL MITCHAM is Professor of Philosophy at Pennsylvania State University, where he also serves as director of the Science, Technology and Society Program. His publications include *Philosophy and Technology* (1972, 1983), *Theology and Technology* (1984), *¿Qúe es la filosofía de la tecnología?* (1989), and *El nuevo mundo de la filosofía y la tecnologia* (1990).

MARGARITA PEÑA is a Colombian historian and educator currently serving as an undersecretary of education for the Colombian Ministry of Education. She has degrees from the Javeriana University of Bogota, the University of Edinburgh, and from Pennsylvania State University, and she has worked for the National Apprenticeship Service of Colombia and at the University of Puerto Rico at Mayagüez. Dr. Peña has written widely on educational policy in relation to technological and social development in Latin America.

MIGUEL A. QUINTANILLA is a professor of logic and philosophy of science at the University of Salamanca. He is the author of *Tecnología: un enfoque filosófico* (1989), *Ideología y ciencia* (1976), and several other books and journal articles in the fields of general philosophy, moral philosophy, and philosophy of science and technology. He has been a senator at the Spanish parliament and the president of the Parliamentary Committee for Science and Technology Policy.

JOSÉ SANMARTÍN is Professor of Philosophy of Science at the University of Valencia. He is president of INVESCIT (Science and Technology Studies Center). He is the author of numerous works on model theory, philosophy of biology, and genetic engineering and society. Among them, he is the author of *Filosofía de la Ciencia* (1983) and the co-editor of *Ciencia, Tecnología y Sociedad* (1990).

ANTONIO E. TEN is Professor of History of Science at the University of Valencia. He is a specialist in the social and institutional history of science in the late eighteenth and early nineteenth centuries and is currently researching the history of unification of weights and measures. Since 1989 he has been director of the Ciudad de la Ciencia y la Tecnología for the Generalitat Valenciana.

RICHARD WORTHINGTON is Associate Professor of Political Science at Rensselaer Polytechnic Institute, and for 1990–1992 is director of the Program in Public Policy Analysis at Pomona College. His published research focuses on the politics of science and technology, and he is currently at work on a book examining the political economy of global production.

MICHAEL E. ZIMMERMAN is Professor of Philosophy at Tulane University and Clinical Professor of Psychology at Tulane University Medical School. Author of *Heidegger's Confrontation with Modernity: Technology, Politics, Art,* Zimmerman has published many essays concerning modern technology. He is currently writing a book on radical environmentalism and postmodernism.

New Worlds,
New Technologies,
New Issues

Part I
New Worlds

Introduction: Philosophy of Technology, or the Quest for a "Hominis ad Hominem ad Machinam Proportio"

GEORGE BUGLIARELLO

TODAY'S NEW WORLDS

As at the time of the great Spanish discovery of America, there are today new worlds that we are beginning to perceive and are seeking to explore. There are also many new technologies that are opening up new vistas, from biotechnology to informatics, telecommunications, and space technologies. This new world and these new technologies are here; they are already a reality. On the other hand, we have not yet developed new philosophical perspectives that correspond to that reality and guide us toward a future in which new explorations and new technologies will continue to develop at a rapid pace. The fact is that we possess today only partial perspectives, and some of those perspectives, if the patterns continue, will likely be proven false in the future. The great challenge to the philosophers of technology is how to connect our new worlds and our new technologies with the search for new philosophical perspectives. The task is a most difficult one, made more difficult by the lack of a substantial dialogue with the technologists, who are usually wary of becoming involved in deep philosophical discussions.[1]

The lack of a clear and balanced philosophical perspective has contributed to the name that some have given to our age, the "age of disaggregation" or, by some others, more optimistically perhaps, the "age of confusion." The Spanish discovery of the "new world" reinforced, through all the development that ensued, the Age of Enlightenment. It led to new perspectives on freedom and to new economic perspectives. The discovery also led to new scientific

35

perspectives. Much of what we see today in the development of science and technology, which at times seems to confuse us, and many of the great political-economic-social changes of today, are the result of that great discovery—of the creation on the American continent of a uniquely powerful and innovative new polity. That polity interacted with the great scientific and philosophical tradition of Europe (a tradition that also owes much to India and the Arab world) and, more recently, with the great technological developments of Japan and other nations in Southeast Asia.

To approach the immense philosophical challenge in front of us—how to develop philosophical perspectives that help us in the intelligent development of new scientific knowledge and new technologies—we need a new intimate combination of philosophy, technology, biology, environmentalism, and sense of society. To do so, I would like to suggest that we must begin with several propositions.

A Sense of Optimism.

The first proposition is a sense of optimism. Never before in the history of humankind has there been greater hope of reducing disease, of achieving better living conditions for every human being, and of achieving universal education. In reality these are more than hopes, as already much progress has been achieved. For example, the self-sufficiency of China and India in terms of food, seemed impossible just fifty years ago. Other examples are the almost total literacy developed in Europe (against the fairly widespread illiteracy before World War I), and a life expectancy now exceeding seventy years in developed countries (versus only slightly more than half that figure in very poor countries). Also, consider the very rapidly developing global consensus on the environment, when less than fifty years ago an innocent ignorance prevailed almost universally.

In spite of this progress, there is clearly a great deal that remains to be done; hunger, poverty, and exploitation still exist. But since 1492 the progress has been immense, even if not always monotonic. A necessary starting point for a credible philosophy of technology is to acknowledge this progress realistically—a progress made possible in large measure by science and technology.

Sweeping Away Clichés and False Ideas

We also need to have the courage to sweep away ingrained ideas that by now have almost become sacred dogmas with some philosophers. Among them is the idea that technology oppresses, that there is a "techno-fascism." How can we feel oppressed if, thanks to technology, we can have food, decent houses, cars, good health care? What oppresses is not technology per se, but the *inequality* of access to the fruits of technology, or the *misuse* of technology, whether deliberate or accidental, whether in the work place (harsh working conditions) or in our social life (invasions of privacy, unresponsive computerized information systems, inflexible standards). We have often seen, rather than oppression by technology, liberation by technology, in the powerful impact of communications technology, defeating even the most undemocratic political systems, or in bringing women to vote thanks to the greater political awareness provided by the presence of radio sets in every home. We have also seen the impact of technology in freeing women from the worst aspects of domestic drudgery, while increasing at least certain opportunities for employment outside the home. A new, powerful technological development is today rapidly coming into being that encourages democracy, diversity, and plurality, while creating a truly global society. That development is a system of human, social, and machine intelligence that can be called "hyperintelligence."[2]

Another such false idea is that there are two cultures, a scientific culture and a humanistic one. This idea was seductively presented by C. P. Snow.[3] In reality, however, we are all living under the aegis of the same technological culture. We all use cars and telephones, watch television, take pills to fix various ailments, and so on. In other words, we all use technology as an extension of our faculties, whether we acknowledge or are conscious of it or not. In reality, we are all living in a subculture of the technological culture, the consumer culture.

A third false idea is that armaments should be eliminated unconditionally. This is a very appealing view, yet we are seldom willing to recognize that because of arms capable of mass destruction, we have enjoyed one of the longest periods in recent times without major wars.

The Need for a New Vision

The philosophy of technology needs to have a sense of realism. Out of this sense of realism must come a new vision of the relation of humans to that most human among biological phenomena, technology.[4] The *hominis ad hominem proportio*—the relation of human to human—that has embodied the classical ideal of our civilization, must now transform itself into a *hominis ad hominem ad machinam proportio,* that is, into an intelligent, compassionate, and equitable relationship of a human being to other human beings and to machines (the machines themselves being, of course, the products of human society).

This new philosophic vision must be a courageous one, as it must encompass a very broad view of biology, society, and technology and be open not only to analysis, but also to the development of normative ethic propositions. If the philosophers do not have the courage to respond to the challenge of a constructive definition of the relation of humans to other humans *and* machines, others will do it, but without the rigor, the discipline, and the rationality that we should expect from philosophy. These philosophical qualities are necessary, for example, to guide us, without destroying democracy, in the transition from a society excessively committed to consumption, to one in better balance with its environment, and with less acquisitive practices. Such a transition, which we eventually must make if we are to have a future on our planet, is an extremely difficult task. It demands a vision of political economy but also—and more fundamentally—a philosophical vision. Unfortunately, religions have abdicated this role, as they do not take into account the machine. A few mass religions consider in a mystical or a compassionate way the relation of humans to other living species, but no religion considers the relation of humans with the machine— that human offspring that poses as many problems for us as our own direct progeny. (Yet, following a Roman tradition, the Catholic Pope is called the pontiff—the builder of bridges, so significant was that artifact to ancient Rome.)

I would like to propose that the foundations of a philosophy of technology—that is of a philosophy that recognizes the human significance of the technological phenomenon—must intimately connect a view of the machine as an extracorporeal product of our own biological evolution, with a view of human beings as biological and social organisms. A *philosophy of technology,* furthermore, should make a major effort, within the framework of this connection, to determine the fundamental tests to which technology must be put to determine its validity.

These tests are necessary because technology, although a product of biological organisms, does not evolve with the slowness and, for most species, the destructiveness that characterizes biological evolution. We must learn, therefore, to distinguish clearly, early in the technological process, between potentially negative consequences of technology and positive ones. Among the former are the imposition of excessive uniformity, insensitivity to human needs such as privacy, destruction of pluralism, environmental damage, or "technofashions" that are an end unto themselves. Positive consequences or impacts are exemplified by the satisfaction of basic needs for food, shelter, health, and employment, and also of less tangible needs, such as the sense of adventure and mysticism that has given us space exploration, cathedrals, and pyramids.

Ultimately, the fundamental philosophical and social question that needs to be addressed is this: What is the relation and the most desirable proportion *(proportio)* of a human to a machine and of society to machines? This, I would like to suggest, is the great challenge to philosophy as we contemplate the new worlds and the new technologies that are opening up to us.

NOTES

1. For a discussion of the limited involvement of engineers in the philosophy of technology, and of the difference in philosophical bent of engineers versus scientists, see my paper, "The Social Function of Engineering: A Current Assessment," in *Engineering as a Social Enterprise*, ed. Hedy E. Sladovich (Washington, D.C.: National Academy Press, 1990), 73–88. Undoubtedly, at least three factors contribute to this limited involvement. In the first place, the education of engineers does not address the philosophy of technology. Secondly, there is the greater complexity of the philosophy of technology, in comparison with the philosophy of science (the former dealing, ultimately, with the very value laden issue of the purpose of modifications of nature and with the essence of the artifact, the latter with the scientific method and the verification of truth). Thirdly, the origin of the philosophy of technology is a much newer subject for scholarly investigation.

2. For a discussion of my concept of "hyperintelligence," see George Bugliarello, "Hyperintelligence: Humankind's Next Evolutionary Step," in *Rethinking the Curriculum Toward an Integrated, Interdisciplinary College Education,* ed. M. E. Clark and S. A. Wawrytko (New York: Greenwood Press, 1990), 25–37.

3. C. P. Snow, *The Two Cultures: and A Second Look* (Cambridge: Cambridge Univ. Press, 1964).

4. It is useful to reiterate that, although some other living organisms produce artifacts (e.g., bird nests, beehives), they do so instinctively, rather than with the deliberately rational and now increasingly science based approaches of humans.

The Postmodern Economy

ALBERT BORGMANN

INTRODUCTION

This essay is an account and a recounting of how the postmodern economy has been rising out of the modern economy. The account is guided by a schema of triads that mark the historical line from the end of the Middle Ages by way of the modern period to the beginning of the postmodern era and its bifurcation into a hyper-modern and a metamodern possibility. The recounting tells the story of how the unified world of the Middle Ages was shattered; how on its ruins the modern project was established; how this project was realized technologically; how modernism, but not technology, has come to its end; how postmodernism is caught in an ambiguous struggle to usher in a new time; and finally, how this ambiguity will be resolved into hypermodernism, if the rule of technology remains unchallenged, and into metamodernism, if we are able to pass through technology to a genuinely other beginning. The essay, then, is both schematic and narrative—schematic to give some order to large and diverse developments, narrative to bring out their moral physiognomy and to end with a moral.

The words *postmodern* and *economy* are not usually joined into one topic. A more common compound is "the postindustrial society," the title of Daniel Bell's book of 1973.[1] I want to speak of the postmodern rather than the postindustrial economy to connect my topic with the wider postmodern debate. I include in this debate not only the authors who have claimed the postmodern label for themselves, but all those whose work falls within the schematic definition of that term. I propose to examine the postmodern economy rather than society for two reasons. First and schematically, my account will concentrate on our management of the material environment; second and narratively, I want to end with the sense of thrifty housekeeping that is conveyed by economy.

THE RISE OF MODERNISM

To begin, I will briefly look at the divide that separates the modern from the medieval period. From where we stand, we can hardly appreciate the grandeur of the medieval world. And yet the achievement of the Middle Ages rests like a shadow of reproach on modernity, and never more so than in the latter's late and closing years. Chivalry and courtesy, community and celebration, authority and crafts are the medieval forms of excellence that are suffering their final dissolution before our very eyes. But the vigor of the medieval order was spent already by the late fifteenth century. Unlike the slow and convulsive decay of Greek and Roman culture, the medieval form of life came to a swift and unambiguous end. It was shattered by the three blows that we commonly associate with Christopher Columbus, Nicolas Copernicus, and Martin Luther.

All three lived at the turn from the fifteenth to the sixteenth century; and jointly, as we can see in retrospect, they shattered the edifice of the medieval world. It was, like all premodern cultures, a locally bounded, cosmically centered, and divinely constituted world. The Columbian discovery of the New World ruptured the finite and surveyable geography of the Middle Ages. The Copernican solar system decentered the earth from its privileged position in the universe. And the Lutheran reformation shook the foundation of divinely established authority.

Again, unlike the fallen empire of Rome, the shattered Middle Ages did not lie in ruins for long. Less than a generation separates the last of the destroyers of the medieval order from the first of the founders of modernity, Francis Bacon. He, along with René Descartes and John Locke, laid the foundations of the modern project. We can speak more specifically still of the three foundational documents of modernism, Bacon's *New Atlantis* (1627), Descartes's *Discourse on Method* (1637), and Locke's *Second Treatise of Civil Government* (1690).[2]

All three treatises are pleas as much as proclamations. They plead for a new order and derive much energy from their indictment of medieval disorder, the duress of daily life, the deadwood of tradition, and the oppression of hierarchy and community. They urge a new fundamental agreement, one that razes the tottering and constricting medieval structures and begins anew on a solid fundament.

Bacon's dissatisfaction was focused on the unprincipled and disorganized state of scientific inquiry. His impatience was fueled by a new understanding of the hardness of human life. Human misery was for him, as it has remained for us, a needless and insufferable scandal that was to be overcome through the domination of nature. Bacon recognized the need for a radically new start, and he was ever eager to fill in the details of the new design. Descartes, to the contrary, argued that the radicality of procedure was crucial. The project of radical reconstruction, he held, would live or die by two requirements. The first was the discovery of an absolutely unshakable foundation. This was to be attained through the fearless clearing away of all existing institutions. The second was a constructive method of irresistible cogency. Once these requirements were met, all problems could be trivially solved. This is the modern triumph of procedure over substance.

Half a century later, Locke drew up a design that spelled out the social implications of the modern project. As Bacon was searching for the "proper foundations" and Descartes was determined to "start again from the very beginning," so Locke was concerned to recast political power by deriving it from its "original," i.e., its fundamental, condition.[3] This he found in the state of nature, governed by reason. There is an element of disguise in the state of nature and the rule of reason, however. These notions lead us to expect an encompassing and stately common order. In fact, however, the *Treatise* is a celebration of the individual, the unemcumbered and autonomous human being. Nature and reason are little more than indistinct backdrops for the individual. The latter is the real fundament of the new social order. The dignity and welfare of the single self are the authority of last appeal. The common order arises from the individuals through an agreement, and this contract remains subservient to the individual.

THE REALIZATION OF MODERNISM

Modernism is the conjunction of Bacon's, Descartes's, and Locke's programs, the fusion of the domination of nature, the primacy of method, and the sovereignty of the individual. It is distinctive of the modern temperament that it has remained unaware and uncritical of the totality of the modern project and of its underlying truth. Two of its components, the domination of nature and the primacy of method, were so deeply internalized by modernity as to become uncontroversial. The great political and cultural debates have been centered on or closely connected to the rights and liberties of the individual. It has, of course, always been recognized that the changing material conditions of modern life came to provide an ever more spacious and commodious stage for the drama of individualism. But never was there a general suspicion that the new physical setting would fundamentally reshape the life of the individual.

We can begin, however, to bring the hidden truth of the entire scheme and story to the fore by giving it a title. The title is technology. Technology in its modern guise and disguise began as the Baconian promise of liberating humanity from the hardness of life and endowing it with the riches of the earth, and it commenced with the Cartesian commitment to achieve liberty and prosperity in a methodical, scientific way. It started in early modern Europe as the rationalization of manufacturing, the expansion of trade, the sophistication of finance, and the systematic exploitation of the conventional energy sources—wind and water—and of the traditional materials—stone, wood, and, within limits, iron.

Expansion within these resources soon ran up against insuperable barriers. Waterflows and waterways are confined to certain locations. Windpower is limited to certain times. Stone is restricted by its heaviness and brittleness, and timber by the slowness of growth. Coal, steam, and iron shattered these barriers in the late eighteenth century and opened up the globe for human domination. Technology in the nineteenth century was massive and often brutal in the reshaping of the material environment. The unbridgeable was bridged by trusses and cantilevers, the remote was reached by railroads, the watery wastes were subdued by steamships, the skies were occupied by highrise buildings.

Technology in the twentieth century represents the refinement and completion of the radical transformations initiated by the Industrial Revolution and the nineteenth century. There were new energy sources, oil, hydroelectric, and nuclear power; new materials, alloys, and plastics. And there were new ways of overpowering time and space through radio, air transportation, and television.

Toward the end of the nineteenth century, the increasing density and complexity of production, transportation, and communication began to threaten the vitality of technology. There was a growing danger that the parts of the technological whole might fail to connect or would work at cross purposes. A new principle of large scale organization was required. In the early twentieth century it was realized in the modern corporation. The modern corporation was distinguished by its gigantic size, its integrating force, and above all by its methodical and hierarchical structure developed and sustained by a new class of professionals, the managers.[4]

One can see, then, how the Baconian program has been realized on a massive scale. The project of modern technology is an eminently real and material affair. It celebrates and affirms reality by overpowering and negating it. It revels in crossing oceans, spanning continents, circling the globe, in disemboweling mountains, plowing up prairies, and piling up cities. The Cartesian design has been carried out in the methodical construction of encompassing organizations, corporations, bureaucracies, and educational institutions.

But what of individualism? It has played an ambiguous role in the drama of modernism. We think of the rugged individual as the modern protagonist. Yet individuals, however rugged, would have been unable by themselves to carry out the Baconian project; and they are entirely submerged in the Cartesian design. The individual can be sovereign only in the realm of commodious consumption as the unencumbered beneficiary of the Baconian and Cartesian machinery.

It is no accident that we think of the modern world as sharply compartmentalized into the Baconian-Cartesian machinery and commodious Lockean individualism, with the machinery constituting a challenging, but culturally uninteresting task and commodious individualism providing the playground of our cultural and intellectual games. It is precisely and only on the structure of a powerful and anonymous machinery that a platform of disburdened commodity and liberal individualism can be erected. But to put matters this way is to reveal the utter dependence of modern individualism as well: there is no playful platform without a supporting structure. It is from this dependence that the modern temperament has studiously looked away.

THE POSTMODERN CRITIQUE OF MODERNISM

Now, as we approach the end of the century, modernism is under attack and about to collapse, or so it appears to the critics of modernism whose favored rallying point is the notion of postmodernism. Since the truth of modernism has for the most part resisted understanding, it may well resist attack and destruction, too. Still, there are concerted attacks on each of its several parts, if not on their truth and peculiar coherence. Moreover, the postmodernist assault shows an impressive diversity of starting points and a remarkable concentration in regards to its targets.

Postmodernism is principally known as an intellectual and architectural movement. In economics there is also a sense of closure and transition, though it is marked with different terminology—postindustrial, postcapitalist, electronic, computer, information. Yet there is, as I will show, a striking convergence of economic, intellectual, and architectural postmodernism.

Each of the three versions of postmodernism has a critical and a constructive side. On the critical side, they are united in rejecting the aggressive realism that comes from Bacon, the methodical universalism that has descended from Descartes, and the liberal individualism that is Locke's bequest. In short, they reject modernism; and though modernism has three aspects, it is good to remember that it is one project and that to shatter one of its sides is to crack the others as well.

Richard Rorty has been the most eloquent critic of intellectually aggressive realism, of the persistent effort of modern philosophers to prescribe both to the world and to everyone within it what may count as real and what may not.[5] Rorty shows us not only that reality as decreed by philosophers is a poor version of the world we know, but also that such decrees owe what force they have not to cogent analysis but to the particular and changeable agreements of communities of speakers. And yet, according to Joan Rothschild and her contributors, aggressive realism is more than an arrogant game of philosophers. It has fueled and sanctioned violence to nature and suppression of women.[6]

Universalism is not only a way of illuminating the modern project scientifically and of coordinating it technologically. It has also been extolled as the lodestar of morality. "Act universally," is what Kant's categorical imperative comes to.[7] Carol Gilligan, however, has taught us to recognize that the individual who has internalized universal principles of justice does not constitute the unquestionable apogee of moral development, but exemplifies a particular and one-sided ethics that has an admirable complement in an ethics of care and compassion.[8] Michael Walzer has given a rich picture of the varieties of justice within a "radically particularist" perspective.[9]

The ambiguity of individualism is laid bare in Louise Erdrich's *Love Medicine,* a portrait of a broken Native American community in North Dakota.[10] All of its members, whether gloomy or gallant, resigned or resourceful, bear the scars inflicted by rugged individuals. At the same time they are kept at the outer margins of commodious individualism. But as Robert Bellah and his coworkers have urged, individualism has been, if less brutal, no less debilitating in the seemingly privileged and protected white middle class of this country.[11]

Regarding, architectural modernism, its paradigm has been the gleaming and functional highrise building that was erected in an open space, cleared of all accretions and confusions of history. It was intended to reorder space and social relations rationally in the face of recalcitrant traditions, climates, and topographies. It was a monument to aggressive realism and the favored target of postmodern architectural criticism. It was attacked for its joyless sterility, its social brutality, and its environmental insensitivity.

So was the grand theory of functionalism and the International Style that sought to impose universally valid principles of design on whatever country and culture. Individualism came to reside in modernist architecture through the anonymity and social isolation of the apartments and condominiums that were aggregated into highrise buildings and through the decline and disappearance of public and communal spaces. Individualism so realized became an object of postmodern analysis as well.[12]

We are now in a position to consider and appreciate the historical depth and resonance of the critical points that have been made by observers of the advanced technological economies. They, too, note the end of aggressive realism. Economically, the domination of nature consists in the exploitation of material resources—timber, ore, coal, oil—and in their transformation into material goods large and small—bridges and railroads, appliances and furniture. For Daniel Bell, extraction is the dominant economic activity of the preindustrial sector, fabrication is that of the industrial sector. Neither sector disappears in the postindustrial economy; both are overshadowed, however, by an activity Bell calls processing.[13] Accordingly, Eric Larson and his coauthors exhort us to recognize this historic change and look "Beyond the Era of Materials."[14] For Bell, the rise and fall of aggressive realism is evident by the succession of the kind of corporation thought to be paradigmatic of the American economy: first U.S. Steel, then General Motors, and finally IBM.[15]

John Kenneth Galbraith depicts a similar passing of power from land to capital and finally to expertise.[16] Land is, as we still say, *real* estate, the most tangible and enduring fundament of wealth. Or so it was until the Columbian discovery opened up seemingly endless resources of land and raised the value of what was needed to appropriate the land, that is, capital. Capital is less real than land. It is relatively mobile and, as financial capital, quite intangible. But the latter is always within hailing distance of the tangible assets to which it constitutes a claim, and its power is dependent on being properly balanced with material goods. Expertise, however, is a source of power, both intangible and self-sufficient. It can generate extraordinary amounts of capital and goods without the favor of unusual material resources, as Japan has well demonstrated.

Cartesian universalism was economically realized, as noted above, in the giant corporations, a handful of which would dominate an entire economic sector. The corporation was intent on stability through universality, absorbing within it, as far as feasible, production from raw materials to retailing, covering the entire market and all but taming the future through planning and capital resources. But this expansive intention has died or disintegrated. General Motors has shrunk, AT&T has been dismembered, and U.S. Steel has lost its very name and identity as a steel producer. Within the surviving corporate giants, their very skeletons of methodical universality, their hierarchical structures, are giving way to flatter hierarchies and more flexible and decentralized organizations.

Individualism, as noted, has led a divided and deceptive life in the modern economy. One part of it is the unencumbered privacy of consumption. In the machinery of the large and anonymous economic enterprises, rugged individualism played its role in sanctioning the unequal distributions of power and prosperity that came in the train of the modern economy. It did so in assuming the mantle of free enterprise and suggesting that, whatever the design of the current social order, it was the outcome of a fair competition among free individuals.

Free enterprise individualism is now under attack, not primarily because individualism as such has been found wanting, nor because competitive individualism has been recognized for the ideological cloak it is, but rather because the economic system it has been covering up is no longer working very well. Free enterprise is an economic system that has had the sanction of the people as consumers, but has been able to conduct its business largely without the direction of the people as citizens. This has worked well enough as long as the national and international economic spaces were relatively open and allowed corporations room to grow and to maneuver. But those times are past, and for the economy to grow as it might, coordination and cooperation are needed. Only the people as citizens can, through the government, act as coordinators of last resort. In this sense, the postindustrial or postmodern society has to become more communal.[17] The call for cooperation has in fact become a cliché of current mainstream economists, though American business finds it hard to heed the call and surrender its sovereignty.

THE POSTMODERN DESIGN

Every worthwhile critique speaks from a constructive position, however implicit. Remarkably, the positive implications of intellectual postmodernism are so deeply enfolded in their criticisms as to be invisible; thus the antagonists of intellectual postmodernism not unreasonably have concluded that postmodern criticism is without merit. But where the postmodern intellectuals have remained implicit to the point of silence, their architectural and economic siblings have been more forthcoming and constructive. They have given the postmodern design some shape and detail.

To the aggressive realism of the modernist style, the postmodern architects have opposed a reverence for tradition and context. They have revived the stylistic vocabulary of classic and romantic building. They take pride not in razing and clearing away the historical context of a city, but in affirming and enhancing it. Rather than developing a grand theory of their own, they work in eclectic and pluralist ways. And, finally, the postmodern architects are concerned to restore and create public spaces that invite individuals to become, once more, members of a community.[18]

Economic postmodernism's critical part most clearly implies its constructive one. Aggressive realism, to start with, has given way to information processing based on knowledge and expertise and requiring relatively little by way of raw materials and resources. The production of goods is yielding to the providing of services. The decline of tangibility and the rise of information has proudly been hailed by David Gelernter: "What iron, steel and reinforced concrete were in the late 19th and early 20th centuries, software is now: the preeminent medium for building new and visionary structures."[19]

The methodical universalism of the giant corporations is under attack by flexible specialization. The paradigmatic unit of the postmodern economy is not the gigantic corporation but the small, expert, and adaptable firm that is able to recognize a market niche and fill it quickly with specialized services or customized goods. Finally, competitive individualism must be subordinated to informed cooperation, if expertise and flexibility are to prosper.[20]

Looking back on the course of history marked by the collapse of the Middle Ages, the proclamation of the modern project, the technological realization of the project, the postmodern critique of modernism, and finally the constructive proposal of postmodernism, one sees a story told by a chorus of diverse speakers and yet exhibiting, so it seems, a remarkably consistent and conclusive schema. But the poorly understood and unresolved status of individualism should warn that the conclusion of the story may be ambiguous and that the ending may still hang in the balance.

The source of the ambiguity, lies, briefly, in the fact that modernism is indeed declining, but technology is as vital as ever. If the power of technology remains unquestioned, modernism will be succeeded by hypermodernism, that is, modernism by other means. If we come to recognize and restrain technology, however, a genuinely other era may dawn, one called "metamodernism" for the time being. The question, then, is whether postmodernism will turn out to be hypermodernism or metamodernism.

To understand the ambiguity, one must understand more clearly the decline of modernism, the character of modern technology, and the interaction of these two phenomena. Modernism is a principled assault on the material environment to disburden and enrich humanity. In a finite environment this kind of enterprise must finally run up against limits, and so it has. There are chiefly three such limits: environmental constraints, goods saturation, and economic crowding.

The first is unequivocally evident in environmental degradation and the depletion of energy resources. Goods saturation is far from being a global problem. All too many countries are suffering from a mortal lack of goods. But in the advanced industrial countries, where the destiny of modernism is played out, the available space for material goods is nearly filled. There is scarcely room for more furniture, appliances, cars, roads, or buildings.[21] The flow of goods has been sustained through more rapid replacement of goods, but there is a limit to this kind of circulation as well.

Economic crowding results when markets are fully occupied and businesses can no longer expand into new resources and customers, but must collide with one another in largely futile and destructive competition. Economic crowding can be ameliorated for a time by removing all remaining barriers of space, time, and culture. This we are in the process of doing. When the process is concluded at the turn of the century, production will shift around the globe in search of the most efficient site, merchandise will pursue every last possible customer, trading will proceed around the clock, and all local traditions will be submerged and suffocated by a flood of cosmopolitan consumption. But meanwhile, economic crowding will grow and imperil economic growth.

Thus the modernist project of aggressive realism and methodical universalism is coming to its end. It is not reaching its fulfillment, however, nor will it simply collapse. It is incapable of either end

because it does not contain its goal. Its underlying truth is technology. The goal of technology, in turn, is first to liberate humans
from the burdens of nature and community and so to make them
over into unencumbered individuals, and second, to shower them
with freely available commodities and so to make them over into
consumers.

HYPERMODERNISM

The pursuit of this goal by way of aggressively transforming
material reality and through vast and hierarchical organizations is
now running up against the barriers of a finite environment. But the
goal of individualist consumption can be retained, and in fact re-
fined, through alternative approaches. Information processing, flexible specialization, and informed cooperation are simply the
technological response to the decline of modernism and constitute
the establishment of a successor culture I call hypermodernism.

Thus, in the face of the planet's material limits, technology is
constructing, through information processing, a sort of dematerialized unreality that we can give the Beaudrillardian name
"hyperreality."[22] Adventures in the hyperreal universe will require
neither travel nor travail. Electronics and media technology will
procure whatever experience and excitement we desire within the
space of our homes. The distant and perhaps unreachable ideal is
the ability to call up whatever persons and events and to interact
with them through the employment of all our senses. Television,
videocassette recorders, and videogames are already transporting
us toward this goal, and the flood of hyperreality is submerging and
eroding reality on a broad front and at innumerable points.[23]

Flexible specialization, as Michael Piore and Charles Sabel have
elaborated, is in many ways a healthy response to the changing
conditions of the late modern economy.[24] But its limits and liabilities are concealed by Piore's and Sabel's failure to consider
either the impact of information processing adequately or the leisure side of the economy in any way at all. Though computer
technology can serve skilled and creative work, it nevertheless
distances one from an engagement with things and persons, and it
does so dramatically when employed in coherent and extensive
applications where the worker's energy is less and less ennobled
and checked by the encounter with real things and persons.[25]

Symmetrically, the purpose of production is enfeebled by the affluence and comfort of consumption. Work no longer responds to clear and present aspirations; instead its products must insinuate themselves into the gaps of consumption modern industry has left. This enervation is aggravated by the service industry. The niches that are being discovered and filled by it were once occupied by the duties and competencies through which we used to appropriate our world and give shape to our lives. Now there is a financial planner to shape your economic life, a color coordinator to select your wardrobe, a wedding adviser to plan your nuptials, and a professional organizer to tell you what of your possessions to keep and what to throw away.[26] With each additional service, a way of being in the real world is cut off and the human person is reduced to an extensionless source of commands.

As the reality of resources and the dignity of wants are evaporating, the character of paradigmatic postmodern work assumes a feverish pointlessness and restlessness that are remarkably like the hyperactivity of distracted and unruly adolescents. Long hours and unconditional dedication are the badge of superiority and authority for those who most decisively extend the frontiers of hyperreality. The labor that is left for the less gifted is dreary and unloved, performed with little ambition and much sullenness.

Computer technology is also furnishing the electronic tissue for informed cooperation. Information is stored and shared through electronic networks of increasing size and complexity. To guard this emerging hyperintelligence from becoming mired in clumsiness and confusion, a strong and encompassing order must be devised and maintained. Accordingly, the individual comes to depend more and more on patterns of communication and stores of information provided by an electronic hyperintelligence.[27]

Critics of the modern economy, especially those who have examined it under the heading of capitalism, have sometimes predicted its headlong self-destruction, because they claim to discern in it a heedlessly aggressive realism and a stubborn adherence to gargantuan and ossified structures. Considering the catastrophic implications of this scenario, one cannot but welcome and admire the resourcefulness of technology at its hypermodern stage. And yet we must acknowledge as well the fatal liabilities of the hypermodern condition, of a life that is enfeebled by hyperreality, fevered by hyperactivity, and disfranchised by hyperintelligence. It is a life that is distended by an acute two sidedness, an enormous increase in our dominion over our experiences and in our sovereignty over the burdens of life, combined with a breathtaking diminution of the human substance. The reduction of the person to an indivisible point of sovereignty constitutes the ultimate and hypermodern stage of individualism.

METAMODERN ECONOMY

But the postmodern condition also holds the possibility that we may recognize technology and pass through it to another beginning, one I have suggested we call metamodernism. What entitles us to that hope is the promise of the wider postmodern conversation. Hypermodern economics responds to the finitude of the real world and to the rule of technology. But not all parties to the postmodern movement have collided with the limits of the environment or complied with the demands of technology. Intellectual postmodernism, because of its cultural orientation, has paid little attention to the problems of ecology and economy. Postmodern architecture, to the extent that it is art rather than engineering, has kept at some distance from technology. Accordingly, the postmodern critique of aggressive realism, methodical universalism, and ambiguous individualism is not inevitably the highway to hypermodernism. It may rather be the rising trail that leads to the divide from which we can begin to glimpse another era. It seems undeniable that we are rising out of the modern era. The fateful question is whether we are climbing up to the joyless and endless plain of hypermodernism or whether we are truly reaching the postmodern divide, a pass that opens up on a metamodern possibility. The path to that possibility is an alternative interpretation of the postmodern critique of modernism. I will attempt to map it out, and my compass will again be the triadic schema that has been guiding me thus far.

When an epoch approaches its end, the norms of that epoch become subject to criticism, and the critique commonly holds those norms to be incorrect or immoral. So it is with the postmodern critique of modernism. Critical zeal springs, understandably, from the extreme effort needed to rise above one's age. But it is essentially inappropriate. An epoch becomes weak and exhausted, sometimes to the point of convulsions, and its norms begin to lose their authority and adequacy. However, they do not become false or insidious. Every epoch contains its own truth. This does not mean that an epoch is by definition beyond reproach. It is to be judged by its truth. Kingship and patriarchy were parts of medieval truth. We can tell who was a good king or patriarch and who was not. Today, however, the question of whether a man is a good monarch is pointless, and the question whether he is a good provider and ruler of his household ought to be. Monarchy and patriarchy are, in the modern world, norms without authority or adequacy, and they become insidious when enforced beyond their epochal life.

These suggestions should not be understood as a metahistorical schema of epochal change. They simply tell us what things look like from the postmodern watershed. Standing on this divide, we have left modern systematic universalism behind and with it all metahistorical schemata. At any rate, I want to affirm the adequacy of aggressive realism as the modern response to the collapse of the Middle Ages, and I want to affirm as well the force of the postmodern critique of Baconian realism. The economic reply is the move to hyperreality. The conclusion that is usually drawn in the intellectual debate is the rejection of realism and the recommendation of pluralist truth by individual speech communities. This conclusion, it seems to me, is concretely realizable only as a collectivized version of commodious individualism. Only at the superficial level of leisure and consumption, supported by a highly coordinated technological machinery, can speech communities cheerfully agree to disagree with one another. Hence, in its constructive suggestion, intellectual postmodernism is tacitly hypermodernist.

Are we then condemned to hyperreality? We are not. The legitimate postmodern problem with the Baconian project is not its realism but its aggressiveness. In fact, aggressiveness is finally the enemy of realism. It sets out to secure reality, but it ends up defeating and even destroying it. The genuine alternative to aggressive realism, hyperreality, and truth by agreement is the careful and respectful realism that is adumbrated in the contextually sensitive and historically reverent side of postmodern architecture. To be sure, the reality that the postmodern architect affirms is not expansive or dominant, but rather like an island in a sea of artificiality. The stately Great Hall of the National Building Museum in Washington, D.C., with its magnificent columns and artfully tiered arcades, or the convivial space of Quincy Market in Boston[28]—these places are real, being visibly rooted in a local and historical context and commanding our attention in their own right, if only we are equal to their eloquence.

Once we are alive to the contextual and the commanding as the marks of the real, we recognize epiphanies of reality beyond the built environment—wilderness areas, rivers, and mountains. They speak to us in various ways. The voice of wilderness is clear and pristine; we hear it with gratitude and perhaps with anxiety about its continued resonance. The voice of the polluted and neglected river is faint and painful; we react with anger and despair, or perhaps with the determination to restore the river to its rightful vitality and sparkle.

Careful attention, then, reveals beneath the overlay of artificiality an underlying reality that comes to the fore here and there. To the hypermodern temperament these appearances are incomprehensible remnants of the days before yesterday. To a metamodern sensibility they are focal points of engagement and orientation, nothing more or less: nothing more, for to strip away entirely the artificial overlay would be irresponsible were it even possible; nothing less, if we allow the revelations of reality to center and grace our lives. In the latter case, our task is to keep artificiality from thickening to the point where it overlies and suffocates all that is real. We must reduce the artificial surface to a marginal and subservient position. It will still be a broad and powerful margin where we spend much of our everyday life. But the final function of the artificial must be to set off the real in its simple and unforethinkable splendor.

To avoid misunderstanding, let me clarify that in the physical sense the artificial is as real as what I have called reality. Physical reality is a necessary condition of the artificial as well as of the particular contextual and commanding reality I will call focal. Focal realism, accordingly, is the devotion to focal reality and the metamodern alternative to hyperrealism.

Turning now to modern methodical universalism, I again accept its rejection in the postmodern debate. Here, I find the ambiguous eclecticism of postmodern architecture too weak a reply and the flexible specialization of postmodern economics too instrumental. The suggestion of pluralism, coming both from the intellectuals and architects, is clearly a characteristic stroke in the signature of any postmodern movement. But if we leave it at a stylized and free-floating pluralism, we again tie the postmodern alternative to a technological substructure and a hypermodern mooring.

Pluralism must be anchored in focal realism and draw its substance from the particular and local things of nature and culture that call forth our loyalty and give us a place in the world. Dedication to such a thing, to the engagement it commands and perhaps to the language wherein it was born, should be the organizing principle of the plurality of cultures. It certainly should not be race, gender, or class, and at length it may not be nationality either.

Local pluralism, then, is the metamodern response to methodical universalism. We are left with the reply to ambiguous individualism. Postmodernists almost unanimously oppose it in the name of some sort of communitarianism. But again, postmodern communitarianism has been vague and thus easy prey to modernist countercharges of patent wistfulness, or latent totalitarianism, and to subversion by hyperintelligence. Postmodern architects have been strenuously at work creating communal spaces. But in reflective moments they have realized that these spaces have in fact been flooded by the individualism of commodity consumption, or they have remained awkwardly empty.[29]

How can we secure substance and a center for communitarianism? The answer is that *we* cannot, nor can the architects among us. Focal things cannot be secured or procured, they can only be discovered, revered, and sustained in a focal practice. Such focal things and practices are well and alive in our artistic, athletic, and religious celebrations. Metamodern communitarianism consists of the courage to heal these communal celebrations of their consumptive infections and to give them a central and generous place in our cities and countries.

Focal realism, local pluralism, and communal celebration constitute the metamodern resolution of postmodern ambiguity. But what kind of economy will be consonant with it? The postmodern economy, characterized by information processing, flexible specialization, and informed cooperation tends, under the rule of technology, toward hypermodernism. But there is as yet enough looseness and uncertainty in the emerging postmodern economy to make it serve metamodern rather than hypermodern ends.

Information processing would then not be pushed ahead to disencumber us from the burdens of reality and persons, but to respond with more insight and respect to the injuries of the environment and to the injustices among people. Flexible specialization would be the response to the possibilities of local, labor intensive industry. Informed cooperation would take the place of oppressive or mindless working conditions.[30]

What unites and centers these several aspects of a desirable postmodern economy is the realization that the modern restlessness of ever searching for an alibi, an elsewhere, must come to an end. We must stop building and begin to dwell in the land that has been given to us. To practice postmodern economy is to put one's house in order, to dwell in it through celebration, and to say: It is good for us to be here.[31]

NOTES

1. Daniel Bell, *The Coming of Post-Industrial Society* (1973; reprint, New York: Basic Books, 1976).

2. Francis Bacon, *The Great Instauration and New Atlantis,* ed. J. Weinberger

(Arlington Heights, Ill.: Harlan Davidson, 1980); René Descartes, *Discourse on Method,* trans. Laurence J. Lafleur (Indianapolis: Bobbs-Merrill, 1956); John Locke, *Treatise on Civil Government and A Letter Concerning Toleration,* ed. Charles L. Sherman (New York: Appleton, 1965).

3. Bacon, *Great Instauration,* 2; Descartes, *Meditations,* trans. Laurence J. Lafleur (Indianapolis: Bobbs-Merrill, 1960), 17; Locke, *Treatise,* 4.

4. See John Kenneth Galbraith, *The New Industrial State,* 2d ed. (Boston: Houghton, 1972), 72–85; Alfred D. Chandler, Jr., *The Visible Hand* (Cambridge: Harvard University Press, 1977); and Michael J. Piore and Charles F. Sabel, *The Second Industrial Divide* (New York: Basic Books, 1984), 49–72.

5. Richard Rorty, *Philosophy and the Mirror of Nature* (Princeton: Princeton University Press, 1979).

6. Joan Rothschild, ed., *Machina Ex Dea* (New York: Pergamon, 1983).

7. Immanuel Kant, *Foundations of the Metaphysics of Morals,* trans. Lewis White Beck (Indianapolis: Bobbs-Merrill, 1959), 22–64.

8. Carol Gilligan, *In a Different Voice* (Cambridge: Harvard University Press, 1982).

9. Michael Walzer, *Spheres of Justice* (New York: Harper, 1983), xiv.

10. Louise Erdrich, *Love Medicine* (Toronto: Bantam, 1984).

11. Robert N. Bellah et al., *Habits of the Heart* (Berkeley: University of California Press, 1985).

12. Charles Jencks, *Post-Modernism* (New York: Rizzoli, 1987).

13. Bell, *Post-Industrial Society,* xii, 26, 116, 126–27.

14. Eric D. Larson, Marc H. Ross, and Robert H. Williams, "Beyond the Era of Materials," *Scientific American* 254 (June 1986): 34–41.

15. Bell, *Post-Industrial Society,* 26.

16. Galbraith, *The New Industrial State,* 45–71.

17. Bell, *Post-Industrial Society,* 128, 159; 160, n. 30; 298, 364, 366; Piore and Sabel, *Second Divide,* 265–67, 269–70, 273–75, 279, 283, 303–6.

18. See Jencks, *Post-Modernism.*

19. David Gelernter, "The Metamorphosis of Information Management," *Scientific American* 261 (August 1989): 66.

20. See Shoshana Zuboff, *In the Age of the Smart Machine* (New York: Basic Books, 1988).

21. Piore and Sabel, *Second Divide,* 184–87.

22. Jean Baudrillard, *Selected Writings,* ed. Mark Poster (Stanford: Stanford University Press, 1988), 2, 6–7, 143–47, 171–87.

23. See my "Artificial Realities," forthcoming in *The Presence of Feeling in Thought,* ed. Bernard Den Ouden and Marcia Moen (New York: Peter Lang).

24. Piore and Sabel, *Second Divide,* 28–35, 258–77.

25. See Zuboff, *Smart Machines.*

26. Annetta Miller and Dody Tsiantar, "The Advice Peddlers," *Newsweek,* 22 May 1989, 60–61.

27. See George Bugliarello, "Toward Hyperintelligence," *Knowledge: Creation, Diffusion, Utilization* 10 (September 1988): 67–89.

28. See Vincent Scully, *American Architecture and Urbanism,* 2d ed. (New York: Holt, 1988), 287–90.

29. Jencks, *Post-Modernism,* 258, 272, 285, 297, 312, 346, 350. Scully, *Architecture,* 290.

30. With the reservations noted in my section on hypermodernism above, I heartily support the proposals of Piore and Sabel and of Zuboff. See nn. 4 and 20.

31. Matt. 17.4 and Mark 9.5.

The Culture of Risk: New Technologies and Old Worlds

MANUEL MEDINA

The most serious problems of our times are creating what I call a "culture of risk." It is evident that the emergence of this culture is closely tied to the development of contemporary technoscience, in part as a direct consequence of that development, in part indirectly, as a consequence of neglect of research to ameliorate the social and environmental impact of technoscientific developments. Nuclear power and weapons plant accidents; chemical industry accidents ranging from manufacturing, as in Bhopal, to almost daily transportation incidents; the continuing contamination of the natural environment and the vital global resources of air and water; the accumulation of toxic wastes without safe means of disposing of them; the deterioration of the ozone layer; the generation of climate change by increasing atmospheric temperature; starvation, poverty, and permanent social crisis in the so-called Third World, which is actually inhabited by the majority of the Earth's population, which continues to squeeze into ever larger, ever more crowded, megalopolises: all of these form part of a long list of natural and social crises of our culture of risk.

The most fundamental crisis, however, is neither natural nor social, but cultural in nature. The societies created by modern European culture have ceded control of technoscientific research, development, and innovation to the technoscientific community itself, including its industrial "arm." The social and political institutions of these societies do not dominate technoscientific production in order to ensure that the latter matches the values embodied in the former. Instead, social institutions are expected to adapt to the requirements of new production schemes, even at the expense of prevailing values. Worse still, modern "developed" societies, in coping with the alarming risks created by new technologies, can only employ outdated sociopolitical assessment mechanisms incorporating dangerously inadequate conceptions of nature, techno-science, society, and their interrelationship.

This is certainly not the first time in history that this situation has occurred. For millenia technological change, either internally generated or imposed from without, has provoked crises in cultures that have led to their transformation, or their eradication. What is distinctive about the current situation is that the risks seem never to have been so great, nor the possibilities for the future so unpredictable. The potential for social transformation latent in contemporary technoscience implicates not only the outward forms of human life, the existing social institution configurations, and our cosmovisions, that is, our visions of the whole into which humans, their institutions, values, and actions fit. Contemporary technoscience also contains the potential for radically refashioning human nature itself, as well as the very nature of the planet on which we live.

Although it has been extraordinarily late in coming, public opinion in some societies is beginning to demonstrate an appreciation of the problems that the indiscriminate application of new technologies poses for nature and for society. With this awareness has come a consciousness of the urgency with which solutions need to be implemented and of the necessity for new approaches if these solutions are to be discovered. Even among politicians there is a growing recognition of the political significance of the social transformations currently being caused by technoscientific developments. Judging by recent events in eastern Europe, it may not be too bold to say that the major international political issues have now decisively shifted from the conduct of the Cold War, now officially over, to global ecological problems, such as transnational pollution and climate change, and global economic problems, both increasingly perceived as tied to technology policies.

With growing public recognition of these technoscience related problems, different societies are rushing off in different directions attempting to resolve them, either by funding technoscientific research on technoscience caused problems, or by changing social and political institutions so as to give the public some voice in the directions of scientific research and technological innovation. The most common approach in modern societies to the solution of technoscientific social and environmental problems is to apply to them scientific knowledge of nature, technology, society, and of their interactions.

In short, the common approach being taken today is to apply science based solutions to technoscience based problems. But if the ultimate ground of these problems lies, in significant part, in our conceptualization of scientific knowledge, then this approach can hardly be successful. And this is indeed the case, for this techno-scientific approach poses serious problems in serving as a basis for justifying social and political action.[1] The principal problem stems, ultimately, from the nature of the prevailing conceptualization of science, including the conceptions of nature and of human society that are incorporated in this conceptualization. Nature is almost always conceived of as separate from society and from the activity of scientific inquiry itself, as an entity with its own properties and laws, which constitute the proper object of scientific investigation and are independent of them. Environmental problems, for example, on this view appear as functional problems within nature's own systems. A solution to these problems would thus follow from a scientific understanding of the natural processes serving these functions. Appropriate technological applications of this knowledge would allow us to correct and control malfunctions in relevant natural subsystems, forcing the processes we desire to occur.

Our relationship with nature, then, is conceived as one in which nature stands over against theoretical contemplation, which is thus quite separate from nature. In such a context, nature can be subordinated to our objective, science based, technological intervention because the two are separate. This situation transposes almost identically to our conception of society: as an independent structure for objective scientific investigation, and as suitable for manipulation based on the knowledge such investigation inevitably yields.

In reality, however, the fundamental relationship between human beings and their environment, whether natural or social, is not a disengaged contemplative one, nor is it a passive theoretical relationship. Instead, this relationship is tied to engaged practices, especially to technical praxis and to human metabolic activity, sociological as well as physiological. What nature means in a given culture derives operationally from the set of technologies available in that culture for interacting with nature.

As with the relationship between human beings and nature, the actual relationship between nature and society is only correctly understood when it is interpreted operationally as a relationship between domains of action, interaction, and mutual consequences. More specifically, an operational understanding of the relationship between nature and society must be based on technologies of interaction with the physical environment employed by peoples, together with the social technologies on which their social organization and interactions are based. But ever since humans developed one particular technical capability, language, each culture has represented, interpreted, and legitimized the framework for its relationship with nature in some idiosyncratic linguistic form, as an aspect of its own parochial cosmovision.[2]

As a matter of fact, conceptions of the origin, structure, and purpose of nature characteristic of each culture are closely tied to technologies available in that culture; but this is obscured by the ideology laden stories about nature and society that are told in that culture. New technologies are therefore always anchored in old worlds, worse, in old fictitious worlds, whose fictitious character obscures the actual grounds of the social and environmental problems inevitably precipitated by the new technologies.

A brief historical account helps explain the various ways that new technologies have been anchored in old worlds. In ancient cultures there were already explicit notions of a relationship between techniques of production and social organization, on the one hand, and symbolic forms of representation and cosmovision, on the other. With the introduction by court scribes and priests of writing, some five thousand years ago, there was a revolutionary transformation of techniques of representation, one that gave rise to a coordinate revolution in forms of social organization. Thus, the technical innovation, writing, underlay the rise of large cities, class structured societies, and centralized empires.

Among the great cosmovisions of that era, that of ancient Babylonian culture is of special interest because of its influence on subsequent Greek, Jewish, and Christian cosmovisions. The Babylonian myths constituted a hierarchical and hegemonic cosmovision that legitimized the social order implicit in state organization and the parallel expansion of the power of the ruling class. The order of nature, commonly on the gods and on the heavenly bodies that governed the city societies, and on the human population of those societies, all dependent on the personalized, superhuman will of a patron god. The divine will was, in this vision, the source of legitimation for forms of social organization as well as for techniques of production that together perpetuated monarchical authority.

In the sixth century, a new cosmovision manifested itself, one that was to become characteristic of Western culture from then on, namely, the theoretical cosmovision of Greek philosophy. In this vision, the individual figures of ancient gods were replaced by abstract entities, and divine action was replaced by theoretical principles. Biological tropes, deriving largely from Aristotle's philosophizing, eventually dominated the language in which this cosmovision was articulated. The concept of nature or physis, for example, was defined so as to stand in contrast to the concept of artifice and artisanal technique. Where a natural object possesses an internal principle determining the course of its development, artifacts are intrinsically inert, the product of external actions, possessing no natures of their own. Consequently, technical knowledge cannot be considered part of the science of nature, not even theories of technique, such as ancient mechanics.

However obvious the distinction between the natural and the artificial may seem, it has been biased from the beginning in a way that isolates the sphere of the artificial, including technological action on the natural, from the sphere of the natural. The biological tropes in which the Greek cosmovision were articulated privileged certain types of techniques as natural, namely, those soft technologies associated with traditional agriculture. Opposed to these were the hard techniques of the artisans. This distinction was not so much theoretical as political, reflecting the power base of the aristocratic landholders who were legitimated in their possession of power by both Plato and Aristotle. Craft techniques, on the other hand, were linked to urban democratic politics to which these philosophers were opposed, and so they excluded craftsmen from political participation. Thus, in spite of formal differences between the philosophies of Plato and Aristotle, both formulated theory centered cosmovisions that, like the mythological cosmovisions of their predecessors, legitimated particular arrangements of production techniques, social organization, and the distribution of political power.

In the Middle Ages, the theory centered cosmovisions of the Classical Greek philosophers were combined with later mythic and religious cosmovisions, resulting in a teleological cosmovision that was replaced by the world view associated with modern science. This world view conceived of nature as a machine, thus uniting the soft technologies of agriculture and the hard technologies of the artisan-mechanics. The new philosophers were philosopher-engineers, among them Galileo and Descartes, who opposed the distinction of the natural and the artificial by identifying the scope of the science of mathematized physics with the domain of nature, identifying theoretical mechanics with natural science (indeed, with the domain of being itself, under some interpretations). Artifacts no longer stood opposed to nature, no longer were devices for tricking or coercing nature into doing our bidding, they were

now expressions of laws of nature. The result was a technomechanical vision of the cosmos, of nature, and of human society.

This modern cosmovision not only anchors the social and political institutionalization of engineering practice and technological
innovation, it also promotes and legitimizes the extrapolation to all
areas of inquiry of the experimental procedures characteristic of
mechanical knowledge production. Modern nature philosophy, articulated so clearly by Francis Bacon, encourages just such a
program of technological generalization, aiming at a uniform scientific practice keyed to a mechanical experimental methodology for
knowledge generation.[3] Scientific research, on this view, takes as
its object only that which will advance human control of nature in a
mechanical way. Scientific research here becomes a form of technological production. Furthermore, it is precisely the union of theorizing with technological production that gave rise to the technoscientific revolution of the second half of the nineteenth century, in
which mechanical motifs in the early modern cosmovision were
supplanted by chemical, energetic, electromagnetic, and nuclear
physical motifs.

The mature modern cosmovision moved from a cold mechanical
universe to a "warm" universe interpreted in terms of ther-
modynamics, electrodynamics, quantum physics, etc.[4] As before,
theories tied to technologies under the metaphor of control were
extrapolated even to theories of cosmic processes that were outside
the realm of technoscientific intervention and manipulation. Nature
and society were presented as governed by laws that, in reality, only
represented the extension to both of ideas of operational control of
devices and processes characteristic of technological intervention.
This ideology of technology as control, and for control, was subli-
mated into an ostensibly ideology free theory of the cosmos.[5] At
the same time, embedding technology in such a sanitized view of
theoretical science perfectly matched claims of authentic progress
that were corollaries of theoretical science and the stories that it
embedded in the modern cosmovision about the origin, development, nature, and destinies of nature, society, and culture.

Even a brief reconstruction of its development, such as this,
testifies to the increasing importance, over the last hundred years
especially, of hard technomechanical action based on deliberate
control and manipulation of natural processes accomplished by
means of whatever artifacts or means will achieve the projected
end. It also testifies to the successful mystification that has accompanied making technomechanical action the paradigm for all human

action, the rationale for which is the putative universality of the theoretical knowledge on which technomechanical action is based, along with the unity of nature. To a considerable extent, many of the current problems attributable to technoscientific development derive from the increasing application of hard forms of action to domains traditionally considered soft, or natural. Agriculture, cattle ranching, and traditional medicine are all examples of such soft technologies that are now firmly within the grasp of technoscience. The result is that all of them have been subjected to intense efforts at conditioning and at directing their characteristic processes in order to achieve some optimal end, where both *optimal* and *end* are defined in terms of parameters of such branches of technoscience as biotechnology, genetic engineering, and molecular biology.[6]

The generalization to all areas of nature, society, and individual human action of methods characteristic of hard technology has not only had important negative consequences for the European culture that gave rise to technoscience; by a kind of cultural colonialism, this same phenomenon has also been exported to other cultures that, ironically, have had their greatest successes precisely with those soft technologies now threatened by the imported technoscience. Many Third World problems derive from this cultural colonization, which is commonly imposed on these societies as a condition of participating in world political and economic affairs. Needless to say, its imposition, sometimes assimilated in quite brutal forms, is utterly indifferent to the autonomous expression and continuing development of autochthonic cultural traditions, values, and institutions.

Global ecological criticisms are often a reaction to the manifest and far-reaching consequences of interpreting as progressive technoscience based interventions in nature and society. These criticisms are sometimes accompanied by calls for a return to premodern biologically based cosmovisions, calls that elicit justifications of the modern cosmovision based on the authority of science. One of the most common reactions of ecological critics to these justifications is a critique of anthropocentricity. Current problems, on this view, are a consequence of making human ends and means a sufficient justification for technological action. What is needed is the replacement of anthropocentric ethical codes with *naturocentric* ones, that is, the recognition of the central value of nature, of natural systems and processes, in place of the central value of humanity and its willfulness. Such naturocentric cosmovisions include some that are linked to scientific theory by their acceptance of an evolutionary biological conceptual framework.

But evolutionary biology comes within the scope of technoscience (molecular biology, genetic engineering, information science), so that these cosmovisions are actually neotechnoscientific, capable of legitimating new hard technologies of intervention while masquerading as ecological criticisms of technoscience. By feeding the myth of absolute technoscientific control by humans of their environment, technoscientific imperialism continues its dominance of modern culture, now through genetic and artificial intelligence research.

If we wish to respond decisively to the many current problems attributable directly or indirectly to technoscience, then it is essential that the technoscientific cosmovision itself be replaced. At a minimum, what is required are new philosophies embodying new conceptualizations of science, of technology, of theory, and of practice, and of their mutual relationship. The intersection of philosophy and technology today clearly precludes any conception of technology, or of science, as merely theoretical enterprises. The identification of science with theoretical knowledge, which is in turn identified with a transhistorical, transcultural, universal Reason, reflects a longstanding philosophical prejudice. In accordance with this prejudice, accounts of science are frequently analytical, couched in terms of assertive discourse and logic, in which conceptual questions of a formal logical kind inevitably take first place. The same theoretical orientation that has led analytical philosophy of science into an academic cul-de-sac is threatening to do the same to analytical approaches to the philosophy of technology. The same sorts of formal logical procedures and methodological analyses are simply transposed from scientific theories and applied to technological innovations. This creates a framework for an account of technology that reduces technology either to a rudimentary form of theoretical knowledge, or to applied science.[7]

But an analytical methodology does not exhaust the ways that theoretical philosophy of science can corrupt a mimetic philosophy of technology. Other transposable attributes include a shortsightedness vis-à-vis the history of technology bordering on blindness, and an aversion to recognizing the relevance of the social contexts of technological action. This stands in sharp contrast to an operational, nontheoretical and praxis based, understanding, whether of science or of technology, that would explicitly be firmly rooted in historicity, in concrete social and environmental contexts, and in the complex factors motivating human activity. The affinities between theoretical philosophy of science and its philosophy of technology counterpart extend beyond academic circles, however. The philosophical exaltation of theory as alone rational, alone authentic knowledge—an exaltation that has at times taken forms close to fetishism—has nurtured a thorough mystification of science. It is this mystification that has made it possible to proclaim, and win wide assent to, the value-neutral character of scientific knowledge.[8] This claim, in turn, has served to legitimate, as perfectly rational, avenues of scientific investigation that entail high risks to human life, indeed to life generally. Even more surprising, however, has been the success of attempting to extend this claim to technology as well.

The claim of the value neutrality of science and of technology because of their theoretical characters is connected with another, less explicit, claim: that philosophy of science and philosophy of technology, too, are neutral in valorative terms. That is, because of their theoretical and objective characters, neither of these are of any use in resolving normative questions or in deciding policies for action. Given that there exists an unbridgeable gulf between theoretical and practical reason, it is useless to attempt to construct any bridge between theoretical understanding and concrete, practical decisions. But such a gulf exists only for one who treats theories as though they had fallen from the sky, while a growing body of evidence, accumulated primarily by historians and sociologists of science and technology, along with a few philosophers, reveals that science and technology are both always firmly anchored in practice.[9] What is needed today, then, is a profound revision in our understanding, not only of science and technology as value laden practical enterprises, but also of nature, society, and their interrelationship. Modern conceptions of these are radically distorted because they have been derived from parochial theoretistic interpretations of reason, science, and technology.[10] In short, what is needed today is an operational philosophy of technology, and of science.

As a first step, an operational philosophy of technology and of science reverses the historical primacy that Western culture has awarded to theory over practice. Operational knowledge, capacities for action, and schemes of action become, for such a philosophy, the primary forms of knowledge upon which all other forms of knowledge, including theoretical knowledge, are constructed. The foundations of such knowledge are not the coherence of theoretical presuppostions or logical sentences, but operational experience and social-historical realities. Operational philosophy of technology is charged with concrete culture, not with abstract theory, and cultural development is driven not by theories, but by operational creativity.

Operational philosophy of technology is, in addition, interdisciplinary through and through. It deals as much with technological aspects of science as with theoretical aspects of technology; as much with theoretical results of investigation as with operational procedures; as much with present developments as with historical antecedents; as much with academic realities as with the social environment; as much with the products and material effects of knowledge applications as with cosmological implications. Without renouncing either critical soundness or competence in making value judgments, operational philosophy of technology is a prototechnique whose objective is a systematic elaboration of the general historical and methodological framework, and of the conceptual apparatus, necessary for an accurate account of science and technology as practiced.

Among the philosophical tasks already referred to that need to be accomplished is a radical revision of our conception of the mutual relationships among science, technology, nature, and society. As has been stated above, the absolute foundation of operational philosophy of technology is a rejection of theoretistic conceptions of reason, knowledge, science, and technology. The technological content of technical knowledge takes pride of place in the operational philosophy, not the logical form of technical knowledge, as in theoretistic philosophy. The distinction is decisive, for the technological content is fundamentally context dependent, social-historical, and value laden, that is to say, operational, as the logical form is not. Furthermore, on the operational view, instead of technology being understood as applied science, scientific theories are understood as theorizing about technological effects and processes, theorizing that reciprocally influences the course of technological development.[11]

The mystification of science achieved by theoretical interpretations of reason and knowledge has, as noted above, been transposed to parallel mystifications of nature and society. Nature thus appears as an independent object of theoretical contemplation ruled by ostensibly objective laws that, in reality, are expressions of a constellation of technological devices, processes, and techniques together with the form of their (social) institutionalization. Attributing primacy to theory entails the sublimation of technological action into scientific knowledge in explaining the cosmos. Simultaneously, it imposes limits on technological action derived from this explanation. That is, the putatively ineluctable laws of nature discovered by theoretical science impose restrictive limits on technology: only those technologies consistent with the laws of nature are possible. In this way, the technoscientific cosmovision stabilizes the potentially destabilizing effects of technological innovation on theorizing, and thus on cosmovision itself, which is fragile because it is based on the misconception that practice derives from theory, rather than that theory derives from practice.

The operational interpretation of nature is based on the underlying technological relationships between people and the world they inhabit, not on theoretical representations of a world. Nature, no less than knowledge, society, and human beings themselves, can thus only be understood historically and locally: relative to an articulated context of human being and action. The relationship of humanity to its natural environment is not at all Platonic, but derives from humanity's metabolic activity—that is, from every kind of building up and tearing down, doing and undoing, transforming and reforming in which human beings engage, in short, from the totality of humanity's technological activity.

The relationship between nature and society, on the operational view, is similarly understood as interactive, reflecting the existence of diverse technological systems wielded as much through various forms of social organization as by groups of individuals. Society, in turn, is understood along soft lines, as the whole complex of social interactions in place at a given time, not as a timeless social form locally realized. The dominance of certain technologies, a contingent not a necessary fact, profoundly influences the emergence of social patterns of organization. Reciprocally, patterns of social organization influence the direction of technological development.

The program thus laid out for the articulation of operational philosophies of technology and science reinforces the observation made earlier in this essay that the central problem of the world today is that of clarifying the mutual relationships of human beings, their societies, and nature. As long as solutions to the multiple crises that envelop us are couched in theoretical terms, rest on theoretical scientific conceptions of reason, knowledge, nature, and society, we will never resolve the paradox of technoscientific progress: that the benefits of technoscience necessarily entail *negative externalities*. While the search for theoretistic solutions is frequently presented in conceptual and moral terms, what is needed are operational solutions acknowledging the praxical (hence technological) roots of human actions, values, and interactions, both with one another and with the world we experience. Theoretical analytical, conceptual, or ethical treatments of crises address symptoms only, while covering up the more basic causes of the symptoms, which derive from errant practices whose adequate analysis is beyond the scope of pure theory.

The impotence of theoretical approaches to technoscientific problems is becoming increasingly obvious and undeniable to the public. Given the privileged status that theory has enjoyed in Western culture for more than two thousand years, it is hardly surprising that the most prominent responses to the recognition of this impotence have been the pursuit of new, more powerful, theoretistic programs. These are dedicated to formulating new theory based versions of science, technology, nature, and society. Such efforts reinforce our long-standing commitment to the unlimited potential of theoretical science and technology as applied science for freeing us from all our ills, but in the end they will all turn out to be fruitless.

What is needed is a thorough, critical revision of our conception of theory itself, but, in accordance with the operational viewpoint, it is not at all clear that such a fundamental change in ideas can take place without accompanying metabolic, that is, technological in the broadest sense, changes as well. In order to create the climate for such changes, it is first necessary to broadcast as widely as possible a convincing case that the entrenched theoretical conceptions really are misconceptions, that theory represents an expression of technological capability rather than the reverse. Ultimately, it is a question of recovering liberty of action, of discovering within the truly inexhaustible operational sphere new perspectives on the technological foundations of social and environmental relationships.

To be sure, the changes that are needed in order to effect this cannot be the object of an intellectual philosophical program only. There must be accompanying political action reflecting fundamental changes in attitudes and values on the part of society as a whole, not merely isolated individuals. The task of an operational philosophy of science and technology is to assist in precipitating these changes, first and foremost by way of an unrelenting critique of theory in Western culture. More particularly, it must offer a critique of the technoscientific cosmovision, relating it to persistent value prejudices in Western culture that have given theoretical notions their privileged status. Concurrently, operational philosophy must trace the development of technoscience as a cosmovision, from the generation of Bacon and Descartes to the present. It must expose this cosmovision as an ideology based on the myths of value-neutral theoretical reason and objective knowledge, myths whose propagation has obscured the social relations of scientific and technological practices, and thus obscured the natures of scientific and technological knowledge as well. Finally, operational philosophy must create a public oriented forum for promoting the pragmatic fertility of operational conceptions in place of the pragmatic sterility of theoretical conceptions, in relation to the many technoscience based problems that press in on us today.

In effect, what humanity confronts today is a political problem, the key to whose solution is the removal of intellectual obfuscations that shape our consciousness of who we are, what we are, what the world is, and what the possible scope of our actions can be on one another and on the world. Entrenched ideas of the primacy of theory and its many corollaries are in part responsible for many of the problems people, societies, and the world as a whole currently face, and they militate against successful solutions to those problems. New ideas, ideas more adequate to the actual nature of human knowledge, especially technical knowledge, are a precondition of resolving these problems that trace back, after all, to technical knowledge itself and to the theory based rationale for its increasingly problematic exploitation. But however central the role of professional intellectuals and educators may be in advancing these new ideas, society's role cannot be reduced to that of passively assimilating some newly defined technological literacy and then voting accordingly.

What is at stake here is the creation of a new world, or at least the acceptance of a new conceptualization of the world, one incorporating a radical renovation of the spheres of technological action, social organization, and cultural values. In the distant past, and even in this century, religious and political ideologies have been vehicles for such transformations. It is an open question whether ideas, apart from their incorporation into such cosmovisions, have played such a role simply by the force of their perceived consistency and their widespread internalization. The operational perspective, however, implies that thinking is always rooted in context-specific action, so that ideas do not exist apart from constellations of other ideas held together by value judgments. Cosmovisions are thus ubiquitous, but they can be more or less adequate to the life situation of a society, more or less expressive of its values, beliefs, and goals.

The technoscientific cosmovision defined a particular world, deriving its power to do so from authentic perennial commitments of Western culture. But the inevitable limitations of that world view are now apparent. Those very commitments are now generating a host of threatening problems and preventing decisive solutions. Technoscience has become systemically problematic and has given rise to the culture of risk. A new world needs to be defined, based on a new cosmovision in which technology and science, nature and society, are conceptualized less problematically. The most important condition for accomplishing this, however daunting it may seem, is the broadest possible involvement of society as a whole in the political decisions associated with such a transformation. That, after all, is the only way that new worlds are created, by their mass perception as having been created.

NOTES

I want to express my appreciation to my fellow contributor Steven Goldman of Lehigh University for his assistance in converting my early drafts, informally translated, into this fluid essay.

1. See, for example, the exchange between Emmanuel Mesthene and John McDermott on this very point in Albert H. Teich, ed., *Technology and the Future,* 5th edition (N.Y.: St. Martin's Press, 1990), 77–125.

2. The reciprocal, mutually determining, interaction between human beings and the world of their experience, especially the initial role of language and poetry as cosmovision-forming technologies, is the central theme of Giambattista Vico's *New Science* (Ithaca: Cornell Univ. Press, 1984).

3. This is, of course, the program outlined in Bacon's *Novum Organum.*

4. Serge Moscovici, *Essai Sur l'Histoire Humaine de la Nature* (Paris: Flammarion, 1960), especially chap. 7.

5. For a more narrowly focused treatment of the motif of control in Western technological history, see James Beniger, *The Control Revolution: Technological and Economic Origins of the Reformation Society* (Cambridge: Harvard Univ. Press, 1986); also Otto Mayr, *Authority, Liberty and Automatic Machinery in Early Modern Europe* (Baltimore: Johns Hopkins Univ. Press, 1986).

6. For one perspective on this phenomenon, see Joan Rothschild, "Engineering Birth: Toward the Perfectibility of Man?" in *Science, Technology and Social Progress,* ed. Steven L. Goldman (Bethlehem, Pa.: Lehigh Univ. Press, 1989).

7. This phenomenon received its first systematic treatment in Jerome R. Ravetz's *Scientific Knowledge and Its Social Problems* (Oxford: Clarendon Press, 1971).

8. A recent discussion of this is Steven L. Goldman's "Philosophy, Engineering and Western Culture," in *Broad and Narrow Interpretations of Philosophy of Technology,* ed. Paul T. Durbin (Dordrecht: Kluwer Academic Publishers, 1990), 125–52.

9. An especially valuable discussion of this is in Bruno Latour's *Science in Action* (Cambridge: Harvard Univ. Press, 1987). See also Wiebe Bijker, Thomas Hughes, and Trevor Pinch, *The Social Construction of Technological Systems* (Cambridge: MIT Press, 1987); and Karin D. Knorr-Cetina and Michael Mulkay, *Science Observed* (London: Sage, 1983).

10. An important discussion of this is found in Ian Hacking, *Representing and*

Intervening (Cambridge: Cambridge Univ. Press, 1983). See also Jerome R. Ravetz, "Ideological Commitments in the Philosophy of Science," in *The Merger of Knowledge with Power: Essays in Critical Science* (London: Mansell, 1990), 180–98.

11. That technology subsumes science is a central theme of John Dewey's work. A recent presentation of relevant aspects of Dewey's philosophy is Larry Hickman, *John Dewey's Pragmatic Technology* (Bloomington: Indiana Univ. Press, 1990).

The New World of New Technology

JOSÉ SANMARTÍN

A number of technical innovations of the past have led several reflective thinkers to the conclusion that they were witnessing the birth of a new civilization. They were convinced that a new era had come, an era that would be more humane than the previous one. The truth of the matter, however, on every such occasion, was that these convictions turned into frustrated hopes.

I do not see myself as one of these optimistic futurists, although I may become one, in the long run, against my own instincts. However, it seems to me that there are certain elements of the current situation that in the past have accompanied moments of authentic cultural crisis. In order to describe these characteristics, it is probably a good idea to start with a brief summary of the elements that I would consider most representative of the *old world*.

TECHNOLOGY AND SCIENTIFIC/INDUSTRIAL CULTURE

I refer, by the term *old world*, to enlightened modernity: the civilization that reached its peak at the end of the eighteenth century—a civilization whose superideology had science as its backbone. In the old world, science was not so much a set of theories as an activity. In fact, modern science was mainly an expression of a pursuit of power and represented the potential to perfect nature.[1]

In the old world, nature was viewed as a sort of enemy. It was said that, in order to perfect nature, one had to do away with this antagonism. Science played a central role in eliminating this opposition. Science made it possible, in principle, to control or dominate those elements in nature that caused us difficulty. Having said this, no one should be surprised if I say that modern science is to a

72

great extent technology and that technology is the most fundamental part of science.

The Superideological Background of Modern Technology

From a superideological point of view, the *old world,* or so-called industrial, technology, has four characteristics that I would like to analyze: progress, technological imperative, technological determinism, and military application.

Progress. The concept of progress emerged hand in hand with modern science. What is expected of technology is the scientific construction of a better world. By a better world we usually mean one in which human well being is continuously increasing. This is achieved, on the one hand, by placing human beings under the protection of modern technology, against the changes that occur in nature. On the other hand, a better world is constructed through the technological reorganization of nature, such that nature provides more and more useful services to humankind. In other words, it can be said that a better world is one that is less threatening from a natural standpoint, and in which the standard of living is higher from a social point of view. The first state would seem to be a necessary condition for the second. Both things also seem to depend on technological innovation—on the identification of new causes upon which to operate technologically and systematically.

For these reasons, technological invention is considered the principal cause of the advances in the control of nature and of the incremental advance of the standard of living. Put simply, technological innovation is the principal cause of progress.

The Technological Imperative. It is quite commonly thought that all technical developments ought to be applied, particularly in industry. This is what we know as the *technological imperative:* we can technically do it, so let us do it. It will work, and if it does not, some other technical development will remedy the situation.

It has been said that statements such as "it will work" or "it will not work" do not have much to do with the direct consequences of the innovation in question. The innovation is considered valuable in itself. Such reference is more to unexpected aspects and secondary effects that could tarnish the technological achievement. The most widely held belief is that nothing is going to turn out so badly that there would be a loss of control of the natural environment or in the standard of living associated with the innovation. And, in any case, no solution of problems caused by a given technology can be found outside technology itself. The negative impacts of a technology can only be solved with a better technology.

Technology and Technological Determinism. The creed to which I have just referred leads to the emergence, in many cases, of an attitude of *technological determinism,* according to which society and human beings are subject to an autonomous technology.

The belief that all technological innovation is automatically beneficial because it contributes to our well-being or to the creation of wealth often underlies the recommendation that technological evolution must not be impeded. Some technological determinists advise that it is not a good idea to oppose technology because of the risk of being swept away by its advance. Others announce the necessity of adapting to, or of having people conform to, technological evolution.

Technology and Defense. The necessity of conforming to technological evolution not only has to do with the conviction that technological innovation, the creation of greater wealth, and the encouragement of more humane social orders are aspects that go hand in hand, it is also associated with the opinion that the safeguarding of these achievements depends on technology itself. This protection takes the form, ultimately, of military defense. Moreover, the innovations effected in the area of military technology are not only considered indispensable for the dissuasion of potential enemies, but it is also assumed that developments in this domain determine the direction of the advance of civilian technology as well. In support of this claim, many examples of the successful application of military technology to civilian contexts are often cited.[2]

The Impacts of Modern Technology

One can detect in all these superideological characteristics of modern technology a kind of justification for the widely held belief that progress entails a certain amount of sacrifice. Specifically, one can try to demonstrate just how pertinent it is to give full credence to the view that we must accept the indirect consequences of modern technology, if we want to enjoy the benefits of its direct consequences: improvements to the standard of living. Thus, if a higher standard of living and well-being is desired, we have no choice but to accept the possible negative effects that may accompany the process of technological innovation—being assured that these negative impacts will be redressed sooner or later by improving the technology that caused them in the first place.[3]

Among the negative impacts of modern technology that, according to the old world of the modern superideology, one ought to accept, are the following:

Technology and Exploitation of Resources. Modern technology uses as sources of energy and as raw materials vast quantities of natural resources that are irreplaceable. These resources have been inherited from what are usually age-old stores of materials that are now rapidly being consumed and, sometimes, squandered. But this practice is, apparently, a necessary one. It seems to be necessary because, in the first place, that intensive exploitation, in particular, the large scale extraction of nonrenewable energy resources, is quite cheap (or at least less expensive than alternative procedures). And this is something that directly affects the price of the technological product. This is a price that, until recently, has not included the cost of repairing the damage caused to nature by the extraction processes, let alone by consumption. The rapid consumption of natural resources seems necessary, secondly, because there is a belief that only natural resources of just these sorts can satisfy the energy requirements of highly concentrated industries and populations.[4]

Technology and Concentration. If there is a single characteristic that has always accompanied the process of technological-industrial development, it has been that of concentration. I believe that the identification of progress with the increases in *efficiency* attributed to technology plays a very important role in the establishment of concentration as the goal of the most significant institutions in modern times. Efficiency is understood, initially, as productive efficiency. Efficient production is defined, in turn, as that which produces the most benefit with the least effort and expense. A larger profit was supposed to provide the necessary basis for a greater and better distribution of wealth; thus progress and the increase in the standard of living are perceived as going hand in hand. But it was also in this manner that progress and uniformity, and progress and concentration, became closely connected processes. In fact, the modern world view, or industrial superideology, is based on the belief that uniformity is a prerequisite for efficiency.[5]

Uniformity invades all aspects of modern life, from ways of dressing to ways of working. This tendency towards homogeneity is linked to the belief that the efficiency of a uniform process increases according to the increase in its concentration. First, the extractive operations tend toward concentration. The success that concentration initially had in the mine induced its transfer to other domains. On the one hand, there was a concentration of petty principalities into centralized states, with one army, one head of state, one language, one set of commercial practices, and so forth.

On the other hand, there was also a concentration of energy sources, with rather disastrous results. Instead of using a variety of energy resources, technological-industrial development depends, more and more, on a very few sources of energy, driving them effectively to depletion.

A similar process occurs in the workplace and in the population. Modern industry tends to concentrate its facilities, and with them the population it uses and serves, in urban centers. The megalopolis emerges, and along with it there emerge problems that range from individual and social pathologies to serious problems of environmental pollution.

Technology and Pollution. The extraction of natural energy resources has been taking its toll on the environment. Moreover, industrial technology had detrimentally affected water, air, and soils. This situation, however, was considered tolerable—and even desirable—because of the greater social wealth that scientific industrial manufacturing was said to produce, by comparison with nonscientific techniques.[6] Thus, traditional farmers turned their backs on their fields, and were lured to industrial work, which promises an easier and more prosperous life-style. In this way, small rural townships are being replaced by major urban centers as the ideal form of existence.

Urban conglomerations generate more pollution than the same population dispersed in scattered villages, but they are said to provide more opportunities for a better life than the villages offer. Thus, to achieve a higher standard of living there seems to be no choice but to bear with the physical pollution and the social pathologies apparently intrinsic to modern technology. This certainly does not mean that improvements to technology should not continue to take place in order to keep this contamination to a minimum. But initially, at least, one must make use of this technology. There will be time, later, for taking care of its undesirable consequences.

THE CRISIS OF OUR TIMES

I do not believe I am exaggerating when I say that this network of impacts—whose crucial nodes I have tried to identify—has expanded greatly since the turn of the century, with the development of nuclear and synthetic technologies.

Nuclear Technology. To the contaminating nature of traditional en-

ergy resource extraction and use of nonrenewable energy resources, one must add the impact, since the 1950s, of nuclear
technology, used both in military and civilian domains. Never before had such potentially hazardous agents been manipulated.
Nothing so dangerous had ever been known, either in the geograph-
ical extent of its effects or with respect to the magnitude of its
potential negative impact on life. Nuclear technology is paradigmatic with respect to what should never have been done. Following
the principle of the concentration of energy sources, and looking at
the threat (later proved to be false) that fossil fuels would become
scarce and eventually become insufficient for supplying the predicted needs of industry by the 1970s, in the 1950s it was quickly
undertaken to build large numbers of nuclear plants. No technologist was unaware of the dangers these plants entailed. Nonetheless these dangers were proclaimed minimal in comparison with the
great benefit that those plants would provide: no other energy
source came near to it in principle, as far as the low cost was
concerned.[7]

Synthetic Technology. Nuclear technology shares a large number of
characteristics with other technologies typical of our times, the
ones I have called *synthetic*.

It is difficult to confuse the results of mechanical technology with
natural elements. The former are not usually expected to imitate
reality. The unity of structure and the function present in natural
objects having been sundered, mechanical technologies aim at reproducing the function without copying the structure. By contrast,
it is often difficult to perceive the artificial origin of the products of
synthetic technology. These products are intended to reproduce
natural structures, or structures that could be taken for natural
ones. A typical example of synthetic technology is a bacterium
obtained through DNA recombination techniques.

Without playing down the risks and impacts of mechanical technology, it is quite clear that those involved in the synthetic domain
are considerably greater. As the technologists with an old world
mentality would say, radioactivity has always been present in
nature, and, ultimately, nothing could prevent nature from producing, with enough time, the chimeras that genetic engineering produces today. I believe this is true, but I also can see a qualitative
change that the risk of synthetic technology implies, as opposed to
the mechanical technology. Synthetic products, added to those
already existing in nature, could cause a serious imbalance in
existing ecosystems, including those central to our survival.

Natural systems, after all, have a tendency to a certain stability.
This natural balance constitutes an important element in the con-

tinuity of life. It is also important to remember that natural systems have a certain flexibility, but not so much so that they can successfully assimilate any modification of its constitution or environment. While it is true, then, that there is radioactivity in nature, that is not to say that any amount of synthetically produced radioactivity can be assimilated by nature without unpredictably disrupting the equilibrium and health of natural systems. Nature's ability to recover depends on the magnitude of the modifications introduced.

Synthetic alterations, today, can be extreme. For the first time in history, we can synthesize living beings capable of growing and multiplying, beings that can be designed such that there are no biotic nor abiotic factors (at least in principle) that can prevent their flourishing in nature.

Undoubtedly, today's dominant technology is not so much the mechanical as the synthetic one—whether it be synthetic chemistry, nuclear technology, or the DNA recombination technique. What is also beginning to be irrefutable is the fact that these synthetic technologies entail serious potential risks. One need not only remember Seveso, Bhopal, Three Mile Island, and Chernobyl to recognize the danger, for the far more frequent contamination caused by chemical-agricultural technologies will remind us just as clearly.

We can try to minimize these risk and impacts, but I believe that, in most cases, such efforts are like throwing to a drowning person a life preserver. And this drowning person is, so to speak, a victim of the shipwreck of the industrial technology and its superideology.

Most analysts would agree that we are witnessing not only the crisis of the industrial technology due to its negative social and environmental consequences, but also a crisis of the super-ideological background that has encouraged those very technological developments. That is, we are witnessing the crisis of the old world.

Progress: A Concept in Trouble. It seems quite obvious that our worsening technological problems are forcing us to question an assumption that has been until now virtually unquestionable: that technological advancement inevitably increases well-being. The empirically correct claim that the progressive development of industrial technology has generated greater and greater wealth must be qualified. Technology does, indeed, generate more wealth as it develops in technical sophistication, but at the expense of irreplaceable resources, of clean air and water, of fertile soils, without leaving the earth much margin for recovery. It is also the case that while industrial technology increases the standard of living for some people in some places, it may at the same time actually contribute to the decline in the quality of life in other areas. In particular these places are often the sources of energy supplies and raw materials or the dumping zones for waste. In short, the concept of progress has entered a period of crisis along with the other aspects of modern technological endeavors among them standardization and concentration.

The Paradoxes of Deconcentration. At the core of this problematic situation that humankind has been experiencing during the last twenty-five years, one can also discern some destandardizing and deconcentrating trends. These trends are fairly evident, for example, in government affairs.

What is happening (at least in Western Europe) in this context is very interesting. Government is exercised as though traditional majorities still existed. However, it is possible that it is the obsolete mechanisms used in public opinion polls, or in general elections, that really creates these results. Before elections are held, it is common to attempt (even subliminally) to make citizens momentarily forget particular issues. Later, these issues turn into a veritable purgatory for the nation's elected government. For the art of governing now—at the end of the twentieth century—has very little to do with the implementation of ideologies, but rather with the solution of problems that concern minorities.

Certain technologies that already exist in our society could help solve the paradoxes I have pointed out. They are technologies that have very little to do with those characteristic of modernity. They are postmodern, not at all giant, and require very little energy consumption. Such developments as microelectronics and, in particular, computer techology are precisely the types of technology that are encouraging deconcentration.

This does not mean, however, that the use of these technologies will make all our troubles disappear at a snap of the fingers. Social changes are also necessary, particularly changes in the mentality of those in power, in order for these technologies to achieve these effects. The first and most important change is to abandon the framework of the modernist superideology when analyzing the impact of postmodern technology. For example, one cannot fully grasp the problems associated with postmodern decentralization from the standpoint of the modern belief that concentration is the key to efficiency.

I hope I will not be accused of an exaggerated technological optimism for having stated this, but I do believe that the use of computer technology can contribute to the development of a new form of democracy more in tune with the trends toward deconcentration to which I have referred. For example, the computer could allow for the effective registering of citizen opinion about matters that concern them, those problems that are usually confined to the fringes when the democratic process is limited to a simple marking of a name on an electoral list. What is shocking today is that the same politicians who say they are promoting the development of postmodern technology are usually the ones who cling to past ideologies of modernity.

The End to the Arms Race. The crisis in contemporary society not only affects civilian matters but the military safeguarding of civilian technological achievements as well. Capitalism and communism alike generally agree that the key to social progress is technological innovation. Their disagreement has not been so much about the worth of the technology, but rather about the ownership of the technical means of production. Both agree not only on this point, but also on the imperative necessity of defending social achievements militarily. Meanwhile, the arms race these blocs initiated after the Second World War, and pursued with vigor for over forty years, seems to have come to an end. At least on the communist side, and perhaps for both, the investment in military technologies did not promote the development of economy-enhancing new civilian technologies, but actually contributed to economic decline. At the same time, new information and telecommunication technologies could not be assimiliated by the secretive, centrally planned communist societies, further impeding development and economic vitality.

To summarize briefly, I would agree with those who believe that we are experiencing an era of crisis. Nevertheless, I do not believe that current changes are merely affecting scattered aspects of society. Rather, I think what is changing is the very core of the superideology that has dominated modernity, namely industrial technology and the preconceptions of progress, efficiency, uniformity, standardization, and military defense, which have been its attendants. Symptoms, such as drugs and fundamentalisms, which are indicative of the crisis in our times, are signs of the contradictions we daily experience. While the concentrated, homogeneous giants of industrial technology still stand, other forms of technology have to emerge and may contribute to a reshaping of life.

THE NEW WORLD OF NEW TECHNOLOGY

Having analyzed the actual crisis of the modern superideology, I would like to offer, not predictions, but rather wishes for what I would consider a better world. When offering future prognoses, those who support the technological imperative usually read the future as if it were a mere extrapolation of the present. For them, tomorrow is a hypertrophied view of today. They fail to see that there could be future changes in the direction of technological development, as indeed there have been in the past.

On Leaders. The technocultural evolution of humanity has not always followed a linear path. There have been numerous—and sometimes spectacular—modifications to its direction. These changes have not only been the result of the discovery and diffusion of new techniques; the emergence of new religious or political conceptions—sometimes embodied in an individual figure—have also played crucial roles. While I do not envision such leaders in today's world, what I do find is a growing belief that industrial technology has had undesired effects, together with the conviction that circumstantial reforms are no longer sufficient to redress those negative impacts, or to avoid new ones.

The Contradictions Between Modern and Postmodern Technology. Some people are convinced that the negative character of these consequences has been accentuated by the effects of postmodern technology. Thus, the coexistence of modern and postmodern technology creates difficulties in addition to those created by modern industrialism itself. While I am among those who believe this is indeed the case, I do not agree with the notion that the way out of this situation is simply to bring down the industrial-technological structure in order to eliminate all obstacles to the rapid diffusion of postmodern technology. Without an adequate social infrastructure, postmodern technology could bring about negative impacts on a par with those of modern technology.

Thus, while I certainly agree that some of these postmodern technologies require less energy than modern ones and can therefore be instrumental to a process of deconcentration, that by itself is not enough to free us from more basic problems that we have inherited from modernity. For that purpose, it is increasingly necessary that we view technology not as an end in itself, but as a means for satisfying socially established goals and objectives, and that we develop postmodern technologies unencumbered by super-ideological views of the past. Thus, such technologies (1) should not abuse unique sources of energy, (2) should not be based solely on the extraction and depletion of natural resources as sources of energy and raw materials, (3) should not undergo the same process of concentration characteristic of modernity.

The following is an illustration of what I am trying to suggest.

There are many who, anchored in the industrial superideology, believe that so-called alternative sources of energy are insufficient for meeting today's demand for electricity, much less those of the future. Those who subscribe to this way of thinking fill the ranks of what I earlier described as believers in the future as a mere extrapolation of the present, with certain hypertrophied features. They are convinced that as modernity has generated industrial and urban giantism, so too will postmodernity simply further increase the size of industry and urban centers. They are also usually incapable of thinking of the simultaneous use of different sources of energy. Instead, even when they imagine and project installations that use renewable energy, they almost always do so following the example of large centralized plants.

In short, I believe that, in order for postmodern technologies to make up, at least in part, for some of the social and environmental damage caused by the *megalomania* of industrialism, they (1) must use the widest range of available energy sources, (2) should use, to the greatest possible extent, self-sustainable and renewable energy sources, and (3) should be of as small a scale as possible.

This is the least that can be expected. A program such as the one I have just outlined is not simply utopian, for not only are such postmodern technologies available, they are much more apt than the technological dinosaurs of the past to comply with the social realities that are beginning to emerge at the end of the century.

Technological Culture. I wish to reiterate once again that it is not only a matter of a change in technology that is required. It is also absolutely imperative to abandon certain deterministic habits. The idea that all technological innovations need not be exploited, the conviction that it is not necessary to subject society to technics, rather technics should be subordinated to the demands of society, must be instilled into the common citizen, into the specialist, whether humanist or technologist-scientist, and into the politician.

It is not at all unusual to hear that the technological sector demands a greater understanding on the part of society. In many cases we are told that the opposition to technical projects by some minority is a result of ignorance or misinformation. If more were known about technology, irrational arguments about technological products, on which the very progress of humanity depends, would disappear. At the same time, it is not at all unusual to hear humanists emphasize the need for more subjects of a scientific-technical nature in the liberal arts curricula. While there is nothing negative, as far as I can see, about either of these calls—for anything that contributes to bridging the gap between the two cultures is positive for the development of humankind—nonetheless a lot more ought to be said and done in this regard.

First, when one speaks of informing the public, one must not confuse this laudable project with what is actually taking place. Public opinion often is not informed, but rather "deformed," in order to gain support for the kind of technological developments that promoters want to promote. The media, academic philosophers of science, and prophets of technological utopia play a crucial role in this process. They often articulate such exaggerated positions that the public ends up adopting attitudes opposite to the ones these groups support.

Secondly, I believe the incorporation of scientific-technical subjects into the humanities curriculum is very appropriate, for those whose primary aim is to reflect upon their own times should not be ignorant of the factors that shape the very structure of the society in which they live. However, this is not enough. What also must change is scientific-technological education. I am not suggesting more literature should be taught in engineering schools. Instead, what is really necessary is a change in the very spirit of education in science and technology to take into account more fully the societal context of scientific and technological practices.

The collapse of the old world will perhaps do away with the belief that technology and science are autonomous processes, unspoiled by the complexity of the physical world or the subjectivity of the psychic one. The shipwreck of the old world may bring about the emergence of a new scientific-technological conscience in which there is no such thing as technology alone: that it must be technology-in-nature and technology-in-society. If this is to be the case however, we must all help bring the new world into existence. Due to the magnitude of the environment and social significance of scientific-technical development, it cannot be left in the hands of scientists and technicians alone. A new kind of politician and humanist, as well as a new kind of citizen, must take active part in the decisions about the social goals and the technologies that they believe appropriate to meet those goals.

NOTES

This paper has been supported by grants PS87-0128 and PB 87-0336 from the Ministerio de Educación y Ciencia, Spain.
1. So says Bacon in *Aphorism III* of his *Novum Organum*.
2. Recent events in commuist countries made clear that this is not the case. Conversely countries like Japan and Sweden, which are less developed from a

military point of view, are among the most developed in civil technology.

3. See, for example, Emanuel G. Mesthene, "The Role of Technology in Society," in *Technology and the Future,* 5th ed., ed. Albert H. Teich (New York: St. Martin's, 1990), 77–99.

4. One could argue the opposite: it is the policy of extraction that lies at the basis of the emergence of concentrated industry and population.

5. This conviction is, in part, the result of the myth of the machine. Remember that according to the industrialist superideology, nothing can be more efficient than a machine, and if anything characterizes the machine, it is uniformity of movements. See, for example, Jacques Ellul, *The Technological Society,* trans. John Wilkinson (New York: Alfred A. Knopf, 1964); and Daniel Boorstin, *The Republic of Technology: Reflections on Our Future Community* (New York: Harper and Row, 1978).

6. It is interesting that in this case, instead of seeing the exploitation of non-renewable resources as one of the main sources of the benefits of modern industrial technology, the cause is found in the scientific character of the technical developments that have occurred from the sixteenth century onwards.

7. The accidents since Windscale and up to the most recent, Vandellós I (in Spain) have indicated that the initial expectations of a cheaper type of energy cannot easily be fulfilled in this case. The principle of "Let's do it now; we can always fix it later" seems to be totally infeasible in the case of nuclear energy. The safety measures that would have to be introduced to make the plants safe would never be entirely satisfactory, and the kilowatt would become extraordinarily costly, so much so, that it would not be at all profitable.

New Technologies/Old Cultures

DON IHDE

A *quincentennial* is a rereading of an event that is necessarily situated in both nostalgia and ambiguity. Today's world is one marked by the emergence of a global, highly technologically textured lifeworld perched—not always comfortably—atop old cultures.

The Columbian voyages from 1492 to 1504 were events that retrospectively are seen as crucial turning points in "our," that is, Euro-American, culture. The opening of the New World appears as a virtual historical anomaly, but one that sets the stage for what was to become the dominance of Euro-American technologized civilization.

The anomalous quality of the event may be seen in both Columbus's own failure to recognize that he had discovered a New World—he died believing that he had voyaged to just off the Orient—and the equal absurdity of his luck, since he predicted a landfall some three thousand miles to the west of Spain. He had believed Paolo Toscanelli's estimate of the earth's size, which placed Japan approximately where the New World turned out to be, against the wiser scholars, who correctly thought the earth to be about eleven thousand miles around.[1] But neither they nor he suspected the existence of the two vast continents of the New World. So this daring, false-belief-situated voyage may be seen as the expansion of the European world at the same time that the Renaissance began to open the intellectual world of Europe to what was to become a science embodied in material technology.

Why should Columbus's voyages be nominated for that watershed? We know today that he was not the first European to discover the new lands. For even if we dismiss the legends of Brendan the Navigator, who in the fifth century may have voyaged as far as Iceland, we do know that Leif Ericson *did* sail to Newfoundland and actually settled for a time amidst the Skraelings at the turn of the millennium.[2]

Yet neither Brendan nor Ericson can lay claim to the establishment of the exchange that became permanent through the voyages
of Columbus. His voyages carry with them the continuity of both
memory and communicative exchange which are culture shaping.
In our history, they were monumental, or so we think as we look
back retrospectively. We see Columbus as the harbinger of, first, the
colonization that took place on the part of Europeans in the New
World and, then, the dynamic, four-century development of both
science and technology, which transformed both European and
American culture and which today has taken the entire globe as its
resource. Brendan and Ericson stand only as episodes and belong
to a fringe of our historical memories.

This cultural current was one that swept over all other cultures.
And even if the two and a half months that it took Columbus to
reach the Caribbean is now spanned routinely in a third of a day, the
event began what is today's Euro-American cultural dominance.

We tend to think that science and technology are distinctly
Western in origin and development, and, insofar as they dominate
the globe today, we could be lulled into thinking of a kind of
colonial victory over the earth. But the outward current carries
under its surface a countercurrent, and if the surface is that of a
dominant new technology, its undercurrent is that of many other
old cultures. What does this countercurrent portend? I shall examine this question by returning to voyages, not just those of Columbus and his European competitors, but of two very different
cultural histories of a more ancient form.

The first group of such voyages lies in a legendary, forgotten
"history," that of the Pacific Islanders. Today we know that virtually every inhabitable island of the Pacific—with its far vaster
reaches of sea than the smaller Atlantic—was reached and settled
by Polynesians, Micronesians, and other islanders long before Leif
Ericson reached Vinland. We know, too, that once settled, these
islanders remained in contact and exchange with one another
through periodic voyages of pilgrimage and trade undertaken in the
same multihulled boats that characterized the distinctive naval
technology of the Pacific. What is amazing about these feats is that
they were undertaken with technologies that were "stone age,"
insofar as they did not employ metals, and by preliterate peoples,
insofar as the cultures were oral ones.

The second group of voyages—as so many of our antecedents
turn out to be historically—was Chinese, and they preceded Columbus by slightly less than a century. These were the voyages of
Cheng Ho from 1405 to 1433 on behalf of the megalomaniac em-

peror Yung Lo. These voyages opened up much of the Pacific and Indian Oceans and reached to the very back door of Europe itself in the Persian Gulf.[3]

As navigational feats and geographic explorations, both of these sets of voyages easily equal that of Columbus. Does our ignorance or disregard of such voyages simply bespeak our own cultural chauvinism? Or does the Columbian venture still remain of greater historical significance?

If we simply assume the superiority of our Western technology, we might be able to conclude at least that neither of the two earlier, spectacular voyages succeeded in spreading either their technologies or their cultures beyond the limits of the originating source. Yet there are ironies here, too. And I shall look at these by first focusing briefly upon the naval technologies of the three sets of sailors and their techniques of long-distance orientation, or navigation.

I begin with the most ancient of the three sets of voyages, those of the Pacific navigators. By Western standards, their craft surely would be judged to be the most primitive of the three. Their "low-tech" multihulls were wooden variants upon hollowed log canoes, stitched together with coconut fiber ropes, caulked and sealed with vegetable tars, and powered by crudely woven materials. Similarly, the instrument fascinated Westerner would be convinced that the totally instrumentless navigational techniques used by the islanders also should be thought primitive.

Such a judgment would be both superficial and chauvinistic. For, in design and function, the variety of ancient multihulls was seen, even by the early European explorers, as having certain superior features to their own slow and hard-to-maneuver vessels. The larger of the sea-going catamarans of the Polynesians were recognized by Captain Cook to be from two to three times as fast as his ships. As they were unballasted, these sailing vessels were also unsinkable. The multihulls, with a relatively simple crab-claw sail, also were more maneuverable than the square-rigged and more complex ships of the Europeans. Indeed, today "high-tech" versions of the multihulls have broken major speed records set by earlier sail craft, including those of the clipper ships, whose waterline length exceeded that of the multihull racers by factors of three or four. What should be historically amazing in this is the failure of the Europeans to adapt this design technology and to turn it into speedy vessels of commerce or war, as they did with so many other technological borrowings from other cultures.

Similarly, the instrumentless navigational techniques of the islanders—only seriously studied by Euro-American anthropologists in the middle to late twentieth century—not only were obviously successful, even in locating new islands out of sight, but were a cultural marvel. These navigators learned to use the stars, wave patterns, bird paths, atmospheric reflections: in short, a most precise and subtle reading of natural phenomena, in an oral perceptual system not exceeded in accuracy by any non-Pacific contemporary navigation schemes.[4]

I shall not go into great detail here, but shall note that, without a compass, the islanders used star motions and parallax phenomena to orient themselves as they moved across the ocean. Swell patterns were used for course direction, and even in storms expert navigators could detect the swell rhythm under the surface confusion. Bird routes bespoke nearby, but unseen, islands, as did the reflection of foliage upon permanent cloud masses about island mountains. Not believing that a perceptual navigation system could be so used, European experts as late as 1957 surmised that old, worked shells with two holes must have been used by Polynesian sailors as primitive wind speed gauges when, in actuality, the two holes were for the thong that made the shell into a pendant for a necklace. The European assumption is that instruments must mediate controlled interactions with nature.

In studying the Chinese navy, sent forth by Emperor Lo, one sees that the disparity with Columbus is even greater. We recall that Columbus had three small ships, the largest of which was the Santa Maria at about 98 feet. Admiral Ho's fleet numbered 317 ships, the smallest of which was the combat ship at 180 feet, and the largest, a *nine-masted* treasure ship of 444 feet! His crews numbered 37,000 men. The ships also were of highly sophisticated design, with waterproofed separate compartments, balanced rudders, easily reefable, junk-type, fully battened sails, and were large and even multistoried.[5]

We know that in navigation Admiral Ho used the compass, which was after all, a Chinese invention, and that he had elaborate charts of the then known world, with compass bearings and the use of other directional instruments. His seven voyages took him as far as Java, Sumatra, all around Indian Ceylon, to the Maldives, and up to the Persian Gulf.[6]

My point so far is simple: naval technology and navigational techniques of the pre-Columbian voyages were at least as good as, and perhaps better than, those of Columbus. The distance and exploratory accomplishments of these voyages were as startling as those of Columbus, or more so. Yet, it still may be claimed that neither of these epoch sets of voyages mark the juncture between an Old World and a New World epoch.

In part, the reason for this lies in the cultural settings that determined the ultimate result of the two non-Columbian voyages. The earliest voyages of the Pacific Islanders, like the ancient pasts of all oral cultures, lie beneath clouded legends. We do not know when they first occurred, only that they succeeded in colonizing previously unpeopled islands. We do not know the actual motivations for doing the exploration. Some surmise that population pressure or tribal wars in which losers had to leave motivated the leap into multithousand-mile voyages. But because the islands were uninhabited and because the voyages did not reach the inhabited mainlands of the Americas, the cross-cultural and cross-material cultural exchanges that followed Columbus did not occur. The voyages were, so far as modernity is concerned, forgotten.

In the Chinese case, the history and motivation is much clearer and even finds some very contemporary echoes. The Emperor Yung Lo, not the first in his history, believed Chinese culture to be superior to all other cultures. His fleet was to express this superiority by traveling to the far corners of the world and, through gift-giving and receiving, to proclaim this superiority. That was one reason why the treasure ship was the largest in his fleet and the combat ship the smallest.

On reaching a port, his ambassadors would march to the local potentate and give elaborate gifts, proclaim China the center of the world, and seek trade and homage treaties, which in many cases they got. Yet, the return in homage and trade was not enough to balance the cost to the treasury of the gifts and of the proclamation of superiority.

By 1433, Admiral Cheng Ho's voyages were ordered stopped, and in another recognizable reversal so typical of China, echoed again in recent events, foreign travel was now forbidden by Emperor Lo, and China turned inward upon itself.[7] To voyage, or even to build a long-distance ship, became a capital offense in the decades following the brief outward movement of Cheng Ho. Thus, while not historically forgotten, the consequences of these Chinese voyages were reversed.

Does this leave Columbus, almost by default, the voyager to have turned the world from old to new? If so, it was clearly not because his outward voyage was one undertaken with either a superior naval technology or superior navigational technique. Of course, one might argue—and this argument has often enough been made—that what differs was the rapacious and hungry culture that Europe had become and within which Columbus's voyages must be located.

Lynn White, jr., the preeminent historian of medieval technology, has argued that it was the West's insatiable hunger for power that transformed its borrowed technologies into instruments that produced work and control over nature.[8] Thus, the Hindu prayer wheel, adapted, became the windmill that drained the marshes and claimed lowlands of Holland. The stirrup adapted from ancient Mongol designs became the platform for the early form of a battle "tank," the mounted knight with horse powered lance for forward thrust.

Daniel Boorstin, who recalls the Chinese voyages, notes with irony that the Western explorers surely would not have been satisfied with ritual recognition of their superiority. The conquistadors who followed in the Columbian wake simply usurped the Native American lands and realms and paved the return to Europe with a trail of confiscated gold.[9]

My point here is not simply to decry the greed and crudeness of our ancestors, although one cannot deny it. My point is, rather, to show that technologies in the ensemble are distinctly embedded into cultures. This is one of the reasons why the previous focus upon the naval and navigational technologies and techniques did not reveal the distinctiveness needed to account for the watershed quality of Columbus's voyage over that of the islanders or Cheng Ho.

What is needed is a deeper insight into the ways in which the ensemble of technologies relates to cultural *gestalten,* particularly those of perspective adopted and of an implicit view of nature. In the case of Columbus, the perspective was set early by the literate culture of the West.[10] Interpretation was largely through seeing the world as text and, in the case of navigation, through reading charts. Portolans, or charts detailing a coastline with dominant wind directions, had been common for two centuries or more prior to Columbus. He also knew the mathematized way of seeing the world embedded in the concepts of latitude and longitude and knew that, if he followed the same latitude westward, he would reach the Indies. But more important, with his mathematical seeing, the crude astrolabes for determining latitude, and, particularly, the ability to assume a fixed overhead position looking down upon the map represented flat earth, he could plot his place upon the spherical earth. I claim this overhead or god's-eye view is deeply embedded in Western culture and is indicative of some of the other features of dominance I have already noted, which belong to the way of seeing typical of Columbus's as of later times.

In deep contrast, the islander navigator did not even conceive of the earth as a globe, but as a "mother" who would bear him to the distant lands in her care. Lacking both the practice of text reading and its overhead view, the Pacific navigator instead followed a practice of learning oral songs that carried the formulae for voyages and a body relativism for calculating apparent directions. The perspective adopted was one of embodied "being-here." Once on board, the navigator takes his actual bodily position as constant and judges all relativistically according to that reference. Even the language changes, and one says, "Tahiti is approaching . . ." rather than "I am nearing Tahiti." There is no bird's-eye view from above that would allow one to measure a distance between sailor and landfall. This embodied way of seeing is consonant with the way of seeing oceanic nature characteristic of that culture.

If there are, indeed, two very different ways in which technologies and techniques are culturally embedded, the question might well arise as to which is superior. There is no simple answer to that question. If we take today's standards and anachronistically apply them, we might be tempted to say that, in terms of accuracy and precision, the Western mode is clearly superior. Today's geosatellites, LORAN, and computer based navigational systems can locate users to within a few feet of their actual position. Such accuracy, however, contrasts just as sharply with the dead-reckoning abilities of Columbus as with the islanders, and it is clearly an unfair standard of comparison.

Nor is it possible to compare fairly the mathematically enhanced and interpreted systems of today's developed technologies with the lack of development of a perceptualistic and relativistic set of possible technologies that could perhaps equally have enhanced the islander mode of navigation. It is only very recently that we have begun to develop what are called *virtual reality* systems, which enhance bodily perceptual capacities to the same degree that our digitalized and mathematized systems do.

Trainers for pilots that simulate actual flying conditions, and fog-erasing displays for actual landings that make runways visible, are preliminary examples of a virtual perceptual system that would enhance a bodily perceptual orientation in space just as well as the already developed system for mathematized, map-reading systems. The fact that one system is better developed today than the other is more an indicator of a sedimented preference of a single tradition than of any absolute technological superiority of one approach to the other.

These very observations, however, point to something much deeper than appears on the surface. They point to a phenomenon of the late twentieth century about a different change relating to new technologies and old cultures.

At the beginning of this itinerary, I hinted that, while it seems obvious to all that the dominant current sweeping the globe in the twentieth century is the spread of Euro-American technological culture, there is no culture, primitive or complex, local or continental, that has not been touched by technologized civilization. Indeed, one of the primary tasks demanded of all cultures is to respond to that current, be it by resistance, modification, hurried acceptance, or supersession.

This process is uneven at present. There are areas where only small parts of modern technology have been adopted, while others have become so modernized that even Europe and the Americas seem doomed to being superseded. (The rush to modernization in parts of the Pacific Rim—such as Korea, Japan, Singapore—is so dramatic that when I return from such places to New York, it seems that I am returning to a decaying civilization.)

I suggested, however, that simultaneous with, and inextricably related to this dominant outward moving current is a countercurrent that returns underneath but along with the top current. That is the countercurrent of the Others. For every contact the Euro-American technologized culture makes with the Other, there returns a countercurrent of the culture contacted. This is the phenomenon of what I shall call *postmodern pluriculture*.[11] It is superficially evidenced in such phenomena as fashion, cuisines, and popular culture. Fashion today is a bricolage phenomenon. It arises not only in the old centers of New York and Paris, but also in Tokyo and Rio. Themes may be Japanese, Spanish, paramilitary, or tribal. But the culture bits and pieces that make up whatever dominant and recessive fashion statements are current are the flotsam and jetsam from the Others of whom we are now aware.

The same may be said of cuisines. Every major capital now has, almost mandatorily, a plethora of cuisines. There will be variants upon Oriental cuisines, both generic and specific Continental cuisines, and some variety of others, often reflecting the trade and taste patterns of a capital. Like fashion, the plurality of cuisines is the backflow from what was the dominant exploratory outflow of Western technological culture.

Popular culture is yet another variant upon the multicultural phenomenon. While jeans may be universal and rock music international and while the elements of that culture are often seen as dominantly American, its roots and development are never simply that. The MTV video is among the most radical of the bricolage phenomena, with its fragments and culture bits the very essence of the short-lived audio-visual presence. Last fall, while I was writing in a very tiny village in northern Italy, the local bar from which I had to make my weekly international call was the center for the local teens to absorb this popular cultural initiation. Here, too, was pluricultural bricolage.

It may be objected that this fragmentary incorporation of multicultural bits is too superficial to characterize the sense of an era. But I have already hinted at a deeper manifestation of the same phenomena. It lies in the comparative study of the three sets of navigators we have just studied. It belongs to the late twentieth century to have analyzed and in some degree appropriated the different sets of principles and techniques that originally belonged to only one culture. Ours is a pluricultural vision. Thus, if today one wishes to maximize speed in sailing, one builds a high-tech version of an ancient Pacific catamaran or trimaran, rigs it with a wing or high-aspect Bermudian sail, and proceeds to exceed the actual wind speed on certain angles of sail. Such a racing vessel owes its concept to a multicultural technological past. The same obtains in the multiple—we say, redundant—navigational systems that we incorporate in flying or submarining or sailing. Today, we alternate from the fixed overhead positioning of the older European map-reading techniques to the relativistic and embodied positioning of the Pacific Islanders, reading nature and LEDs.

This brocilage, multiple seeing, is a seeing that is more distinctively a development of the late twentieth century. It is the inheritor of our practices—if not yet of our believable ideology—of cultural relativism. It is the seeing with what I call the compound eye. That vision is never single. It is a series of multi- and alternative visions, symbolized in the growing presence of the multivisual screens we have become familiar with, such as those found in all television newsrooms, in control centers for large plants, and even in department stores. Compound vision is multiple vision. One scans the multiple screens, focusing here, then there and, out of the mélange, forming new directions and possibilities.

What we perhaps have not noticed is that this new vision, relativistic and compound, is both Western and yet post-Western. It is Western in that it arises and takes place in the matrix of a Western-

originated technological culture, but it is not Western in that what
have been the foundational and dominant ideologies we have taken
for granted now find themselves placed alongside other, distinctly
non-Western beliefs and ideologies. This is frightening to some, but
exhilarating to others.

Of course, it is not the case that such multicultural phenomena
are entirely new. Spain's history, for example, is a witness to pluriculturality, at least in the sense that it became what it was by the
combining and blending of Moorish, Jewish, and Catholic cultures.
History perhaps always has been driven by multiculturality, but our
contemporary compound vision enhances our appreciation of this
plurality and raises it to a prominence not previously noted. But it
also, in its relativistic tolerance, refuses to impose or colonize that
which is Other in its midst. At least, that is the secret hope lying
within a compound vision.

So, just as the Jewish mathematicians located in the school of
Henry the Navigator, the revival of ancient Greek science through
the Arab philosophers, and the universalism of the Christian religion became the basis of what was to become the world march
issued from the Columbian voyages, so today we stand at the
threshold of another new world, this time one that is the globe itself.

The shift in vision, however, is not one that belongs solely to a
Western evolution. It is one that, while linking the globe into a
single communications network, incorporates a diversity it has yet
fully to recognize, let alone deal with in a synthesized fashion.
Cultural terrorism takes its place alongside enlightenment-toler-
ance cultures, socialism finds itself accommodating to consum-
erism, and ancient religious strife takes distinctly nationalistic
shape. The vision of the compound way of seeing remains, at the
dawn of the twenty-first century, at best a bricolage of the pluricultural. But its world is a new world, newer and as different from
the modern one it is now leaving as Columbus's Indies were from
the Orient he had sought.

None of this is to say that pluriculturality in our technological
civilization is without its own distinctive problems. But pluriculturality is what gives these problems their distinctive set. Indeed, many of the most problematic of our social-political problems
may be seen as a response—whether utopian, negative, ambivalent,
or resistant—to technological change; and as an enhanced sense of
ethnicity and even ethnic conflict that inevitably accompanies this
awareness.

For example, with the breakdown of previous socialist versus
capitalist ideologies apparent between Eastern Europe and the

West, there has emerged—or reemerged—a series of struggles concerning ethnic groups and decentralized regionalisms. Similarly, much of the intense strife within the Middle East simultaneously displays revived senses of religious fervor (Muslim fundamentalism, though all fundamentalisms display similar characteristics) in forms that react both to technologization and to apparent threats to traditional religious and ethnic values.

Postmodern pluriculture is essentially not neutral. The plural view is acidic in its effect upon all deep monocultural traditions, and clearly this is deeply felt within monolithic ethnic cultures and religions. Here the conflict of new technologies and old cultures appears as one of the distinctive problems of postmodernity.

It is too early to tell what the outcome may be. One may hope that a dystopian result, conflagration around new issues postdating the previous superpower conflicts, can be avoided. Yet, any simply utopian hope, like the ultimate resolution of religious intolerance resulting from the Western enlightenment, seems more difficult in today's large mélange. But like the secret acidity of the enlightenment, which solved the problem of religious tolerance only by making religions redraw the lines of truth from absoute and public at least to absolute and private, so may pluriculture have to hope for the same aim. This is to say that pluriculture also changes culture, and that will be, in one way or the other, the result of the meeting of new technologies and old cultures.

NOTES

1. Daniel J. Boorstin, *The Discoverers: A History of Man's Search to Know His World and Himself* (New York: Vintage Books, 1985), 224–31.

2. Ibid., 209–17.

3. Ibid., 4.

4. Pacific navigation techniques have been well documented and described by Thomas Gladwyn, *East is a Big Bird* (Cambridge: Harvard Univ. Press, 1970), and David Lewis, *We, the Navigators* (Honolulu: Univ. of Hawaii Press, 1972). My text condenses some of these findings.

5. Boorstin, *The Discoverers,* 190.

6. Ibid., 191.

7. Ibid., 199.

8. Lynn White, jr., "Cultural Climates and Technological Advance in the Middle Ages," *Viator* 2 (1971), 177ff.

9. Boorstin, *The Discoverers,* 190.

10. Don Ihde, *Technology and the Lifeworld: From Garden to Earth* (Bloomington: Indiana Univ. Press, 1990), 66–69.

11. Ibid., 164ff.

Part 2
New Technologies and
Political Responses to Them

Introduction: Technological and Cultural Change—Past, Present, and Future

MELVIN KRANZBERG

History deals with change, and the relationship between human history and technological change goes back to the very beginning of our species. Tools served as extensions of an individual's hand and amplifiers of muscle power, enabling one to adjust hereditary organic equipment to an almost infinite number of operations in virtually any environment, from the ocean's depths to the near vacuum of space. Modern physiology, psychology, evolutionary biology, and anthropology all combine to demonstrate to us that Homo sapiens (humans the thinkers) cannot be distinguished from Homo faber (humans the makers). Indeed, we now realize that humanity could not have become thinkers had we not at the same time been makers. Hence, technology is viewed as the earliest and most basic of human cultural characteristics.

Technology also played a major role in creating civilization. Civilization began, Lévi-Strauss and Childe argued, when humans cooked their food and with the development of settled communities—and both rested upon technological innovations: controlled fires and agriculture. Unlike the hunter, the agriculturist had to cooperate not only with nature but with other human beings. With the coming of agriculture, therefore, came humanity's dawning awareness that we must live and work together with others in order to survive—and that was the beginning of civilization.[1]

For a very long time technological change was slow. But beginning in the eighteenth century, the Industrial Revolution and a concomitant agricultural revolution changed all that. The factory became the workplace, the city the dwelling place. New social classes emerged. The nature of the family changed. The educational role of the family—and the church—changed, as the state took on the responsibilities of schooling, while other institutions began taking over other social activates. Changes occurred in the workplace too. Industrial work required new skills, and the worker's relation to the task shifted. Finally, there was a psychological transformation, namely, our heightened confidence in our ability to use resources and master nature.

Science and technology began to dominate human thought and action in the same way that religion and the changing of the seasons had done in earlier times. Thus industrialism transformed—revolutionized—human insitutions, thoughts, and actions. It changed where and how people lived, worked, thought, played, and prayed, and brought into being virtually all the characteristic elements of today's world.

Despite the role of technology in the vast panorama of history and its continuing transformation of human lives, many humanists still regard technology as something divorced from the essence of humanity. What repels them about technology is what they term the *inhumanity* of its objects, for example, the industrial robots that they fear will make people expendable, or the *anti-humanity* of its byproducts, such as the pollution threatening the global environment. They mistakenly think of technology as merely mechanical, but machines are not something apart from humankind. Technology is about human work, hopes, and desires; it extends to every domain of human activity, influencing the beliefs, habits, and assumptions upon which people base their lives. And technology's significance lies in what it does, for the function of all technology is use, and at the same time, alas, misuse and abuse.

With the onset of industrialism, the question of how technological developments interact with the sociocultural milieu came to the fore. My own reaction has been to formulate the uncertainties involved in the technology-society interaction into what my students have called Kranzberg's First Law: "Technology is neither good nor bad, nor is it neutral." By that I mean that technology's interaction with the social ecology is such that technical developments frequently have environmental, social, and human consequences that go far beyond the immediate purposes of the technical devices and practices themselves, and, indeed, the same technology can have quite different impacts when introduced into different contexts or under difference circumstances.

In judging the efficacy and value of technological development, then, we must also take cognizance of varying social contexts as well as compare short-range and long-range impacts. While nineteenth-century romantic writers and social critics condemned the incoming industrial technology for the harsh conditions under which the mill workers and coal miners labored, G. Duby's careful studies have shown that conditions on the medieval manor were even worse.[2] Similarly, as economic historians point out, although the early factory workers labored long and hard, in the long run their living standards improved as industrialization brought forth a torrent of ever-cheaper goods that were made available to a growing public.[3]

Nevertheless, the impact of technology upon the social and natural ecology makes many sensitive individuals fear that this very human activity—technology—has grown so large and has presented humankind with such awful byproducts that it threatens to engulf us: war, environmental deterioration, extinction of species, pollution, destabilization of ecologies. But if modern technology is so harmful to humans and nature, how do we account for the fact that the standard of living is higher in the industrialized portions of the world than in those where technology lags behind? Why are the citizens of the industrial nations better fed and longer lived? Why is more being done to protect the environment and ecology in the technologically advanced nations than in those which retain a primitive, preindustrial technology?

Why is it that, with but few exceptions, the most technologically developed countries are those that enjoy the greatest amount of democratic freedom, have eradicated cruel and degrading punishments, uphold religious freedom, and, in short, endorse fundamental human rights? Why are the most industrially advanced states the ones that have provided for equality of the sexes, abolished child labor, condemned racial discrimination, recognized the right of workers to associate, developed social security systems for the aged, and, in short, upheld the concepts and practices of social injustice? Can all this be ascribed to coincidence?

I do not mean to imply that modern technology is freeing humankind from all its problems. Many of the dangers that threaten our world arise from misuse and overuse of technology, deriving often from greed and hunger for power and sometimes from ignorance of the long-range impacts of technical application. When a technology brings problems in its wake, all too often we attempt to ameliorate the situation by a "technological fix," that is, by the application of more or better technology, which again creates new problems. Although we can point to scores of technical advances that have made life easier and more pleasant in the industrial nations, the litany of the many problems besetting us offers ample proof that technological advancement by itself is no guarantee of peace and plenty, happiness and prosperity. Yet, while technology by itself might not provide us with utopia, there is no indication that turning away from technology would do so. Indeed, does anyone seriously believe that we can meet the future food, water, and material needs of humankind by cursing technology? Can one truly object to encouraging scientific technology so as to meet the energy, material, and food needs of tomorrow's world?

But this does not mean that a group of technocrats, or a scientific and technological elite, should alone make decisions for the future directions of technology. For while it is true that many of the problems facing humankind today and tomorrow involve technology, they also involve human values, social organization, environmental concerns, economic resources, political decisions, and a host of other sociocultural factors. These are interface problems, that is, problems lying at the interface between technology and society, and they can only be solved—if they can be solved at all—by weighing complex social considerations.

Practical issues of knowledge and education are involved in such considerations, as they are in matters regarding the sociocultural, political, and economic implications of technological change. The complex problems raised by today's and tomorrow's technology require a diversity of knowledge and experience that is far beyond the capability of any one individual or single field of study if they are to be resolved. Thus, just as scientific/technological research and development has become a team effort, so taking into account the human, environmental, and sociocultural implications of technical advances must be a joint effort of scientists, engineers, social scientists, and humanists, politicians, industrialists, and citizens from all walks of life.

Our new technologies have indeed created a new world and raised new issues regarding the interactions of technical advances with social, cultural, economic, political, and human factors of all kinds. I call your attention to the need to link these technological developments to the full range of Western culture's intellectual, spiritual, social, and artistic aspirations. In this way, the newest technologies might enable us to fulfill our oldest hopes for the future of humankind.

NOTES

1. V. Gordon Childe, *Man Makes Himself* (New York: New American Library, 1951); and Claude Lévi-Strauss, *The Raw and the Cooked,* trans. John and Doreen Weightman (New York: Harper & Row, 1969).

2. Georges Duby, *Rural Economy and Country Life in the Medieval West,* trans. Cynthia Postan (Columbia: Univ. South Carolina Press, 1968); and *The Three Orders: Feudal Society Imagined,* trans. Arthur Goldhammer (Chicago: Univ. Chicago Press, 1980).

3. T. S. Ashton, *The Industrial Revolution, 1760–1830* (Oxford: Oxford Univ. Press, 1948).

Culture and Technical
Responsibility

PAUL T. DURBIN

> What I relate is the history of the next two centuries. I
> describe what is coming, what can no longer come differently:
> the advent of nihilism. . . . For some time now our whole Euro-
> pean culture has been moving as toward a catastrophe, with a
> tortured tension that is growing from decade to decade: rest-
> lessly, violently, headlong, like a river that wants to reach the
> end, that no longer reflects, that is afraid to reflect.
> —Friedrich Nietzsche, *The Will To Power* (as
> quoted in Daniel Bell, *The Cultural Contradictions of
> Capitalism*)

I begin with Daniel Bell quoting Friedrich Nietzsche: the modern
world, with technology as its driving force, has reached its culmination in cultural nihilism. I next turn to Herbert Marcuse who, after a
brief glimmer of hope based on New Left activism in the 1960s and
the then new radical feminist movement in the 1970s, quickly
returned to his old despair of activism. Marcuse based his last book
on the premise that the only culture left is that of "bourgeois art,"
left over from the nineteenth century in museums and similar
institutions, the radical implications of which no one in our one-
dimensional age can any longer appreciate. Third, I mention an
equally pessimistic Jacques Ellul who can be interpreted as saying
that traditional religion is our only hope against a deterministic
"totality of technique." Along the way, I admit the nihilism of much
modern art, and with others I lament the very real threat to traditional cultures, for example, in the Third World, that technological
progress represents. Yet, in the end, I am optimistic, and I turn to
recent interpretations of John Dewey that offer hope that we can
create new meaningful symbols, a new culture for our troubled
technological world. I think we can do this, however, only if we
struggle to do so.

My concern here with technology and culture is part of a larger project in which I attempt to deal with the social responsibilities of scientists, engineers, medical researchers, and other technical experts. I believe that, to exercise these responsibilities, such people must work alongside other activists trying to make ours a better world. For me, a better world is not just a world in which the great social problems of the day have been solved. A better world is one in which as many human beings as possible find a fuller meaning for their lives—find new cultural symbols that celebrate the significance of new accomplishments at the same time that they motivate the communities in which we live to continue the perpetual struggle to seek after better worlds.

Before beginning, I want to say a word about what I mean by *culture*. In general, I take culture to mean those parts of social behavior that have to do with what makes life worthwhile in any social group; that is, culture refers not just to any social activities, but to those that have an enhancing quality. More specifically, these aspects of social behavior tend to be institutionalized as the whole range of the fine arts (popular arts tend to get included only insofar as they represent something of enduring value), higher learning and its associated professions (law, for instance, tends to be included not in its mundane but in its more philosophical dimensions), education more broadly, and such leisure pursuits as tourism and the enjoyment of nature. In my view, furthermore, religion and religious institutions definitely fall within the scope of culture insofar as they claim to provide a spiritual meaning for life.

A second preliminary note: the number of authors who write about technology and culture is surprisingly small by contrast with those who write about technology and society. I concentrate here on a handful of authors who are especially well known for their contributions to discussions of the topic. And I begin with Daniel Bell, a social theorist better known for two earlier books—*The End of Ideology* (1960)[1] and *The Coming of Post Industrial Society* (1973)[2]—than for the book I focus on here, *The Cultural Contradictions of Capitalism* (1976).[3]

DANIEL BELL ON THE NIHILISM OF MODERNISM UNDER TECHNOCAPITALISM

I begin with Daniel Bell for two reasons. First, because he begins with Friedrich Nietzsche who has become an international symbol of the nihilism of modern culture.[4] But I also begin with Bell because he has a comprehensive view of the meaning of culture. He includes under that term not only the fine arts, but also religion and anything else that gives meaning to the life of a society in a particular historical epoch.

According to Bell, Nietzsche's is but one of two nihilistic views of our technological age. The other author he refers to is Joseph Conrad, especially in Conrad's novel *The Secret Agent* (1953).[5] There, Bell notes, Conrad has the first secretary of the Russian embassy plot an attack that "must have all the shocking senselessness of gratuitous blasphemy"; it "must be against learning—science." Bell summarizes for us, "The act is to blow up the Greenwich Observatory, the First Meridian, the demarcation of time zones—the destruction of time and, symbolically, of history as well."

It is against this backdrop that Bell sets his project in *The Cultural Contradictions of Capitalism:*

> Is this our fate—nihilism as the logic of technological rationality or nihilism as the end product of the cultural impulses to strike down all conventions? . . . I wish to reject these seductive, and simple, formulations, and propose instead a more complex and empirically testable sociological argument.

Bell's argument, he says, "stands in a dialectical relation with" his earlier book, *The Coming of Post-Industrial Society.* Each concentrates on the structure of postmodern society, with one emphasizing the world of work, the other, leisure activities.

Bell's sociological analysis opposes holistic views of historical development as well as both functionalists (he lists Emile Durkheim[6] and Talcott Parsons[7]) and Marxists. He accuses both of having "unified" systems (a common value system or a determinate culture/economics dialectic). His view distinguishes, as an analytical framework, three axes:

1. The techno-economic order, the "axial principle" of which is functional rationality, with utility as the basic value, and whose structure is bureaucratic and hierarchical

2. The polity, whose principle is legitimacy, and in democracies that means the consent of the governed, with equality as the basic value in all spheres, and whose structure is participation *via* representation

3. Culture, about which Bell, following Ernst Cassirer, says there is no unambiguous principle of change and little structure today in this realm of expressive symbolism, although it manifests a peren-

nial return to basic existential concerns of loyalty, love, tragedy, death, and the basic value of a meaningful life (See pp. 10–12.)

As between these axes, there are different rhythms of change, and there is no determinate relation of one to another. (This is clearly opposed to Parsons—see the reference above—and to Marx, discussed here by way of his neo-Marxist disciple, Marcuse.) "Sociological history," in Bell's hands, traces the "disjunction between social structure and culture," with a primary focus on a "modernism" of "rampant individualism" especially in the cultural realm (pp. 14 and 18–20). Indeed, the book, though it frequently replays the expert bureaucracy themes of *The Coming of Post-Industrial Society,* is in large part a documentation of the anarchy of cultural modernism in the twentieth century—and increasingly in recent postmodernist decades.

Bell's basic point may perhaps be captured in a contrast:

> In the past, human societies have been prepared for calamity by the anchorages that were rooted in experience yet provided some transtemporal conception of reality. Traditionally, this anchorage was religion. . . . Modern societies have substituted utopia for religion— utopia not as a transcendental ideal, but one to be realized through history (progress, rationality, science) with the nutrients of technology and the midwifery of revolution.
>
> The real problem of *modernity* is the problem of belief. To use an unfashionable term, it is a spiritual crisis. (28)

To his credit, Bell admits that religion—though it might restore "the continuity of generations"—"cannot be socially manufactured"; neither can a "cultural revolution be engineered" (30). Cultural values, whether religious or artistic, arise spontaneously, not on governmental or corporate demand.

This is an impressive complement to the impressive sociological analysis of contemporary expertism and technological bureaucracy that Bell had provided earlier. (The view is now commonplace and had been elaborated earlier by prescient authors such as John Kenneth Galbraith.[8] I agree that there is a great deal of anarchy in the cultural realm today. Nonetheless, I think there are problems with Bell's analysis. For one thing, though he blanches at being called a "neo-conservative" (xi), he invites such charges when he says, over and over again, that he "respects tradition" or "authoritative judgments" about the qualities of works of art. (See p. xv, although such claims pepper the text.) And I have already noted his high regard for traditional religion. What I think he means in rejecting the neoconservative label is that he does not want to return to the past; he wants new values to emerge in our own time—something with which I agree. Yet he gives very little guidance on how to go about discovering such new values.

Second, Bell uses the term *liberal* in much too homogeneous a way. I claim that liberalism has a progressive wing that Bell almost never mentions—or else he disparages what he takes to be the excessive egalitarianism of its policies. Perhaps burned by the New Left activism he so strenuously opposed, Bell can recognize little value in progressive liberal activism, and he seems extremely pessimistic about our cultural crisis today.

One might wonder if someone at the opposite end of the political spectrum might not be more optimistic. But radicals of the Left do not often talk about culture in Bell's broad sense, and those who do, when they discuss what passes for culture today, are as pessimistic as Bell. This is especially true of Herbert Marcuse, generally acknowledged as "the philosopher of the New Left" (at least in the U.S.).[9]

MARCUSE ON "THE AESTHETIC DIMENSION"

In his excellent intellectual biography of Marcuse,[10] Morton Schoolman notes how Marcuse for a short time continued, even after the decline of the New Left in the early 1970s in the U.S., to argue that "liberal strategies are best suited to push toward the realization of revolutionary goals" (325). Marcuse even picked out another change agent to replace the now quiescent students. He said the women's liberation movement is "perhaps the most important and potentially the most radical movement that we have" (quoted by Schoolman, 325).

However, even an optimistic Marcuse—according to Schoolman—eventually became disillusioned. In his last book, *The Aesthetic Dimension* (1978, Marcuse died in 1979), Marcuse, says Schoolman, "departs from the optimism of *Counterrevolution and Revolt* [1972] and returns to the theory of technological domination and one-dimensional society" (346). In Marcuse's words: "In the present, the subject to which authentic art appeals is socially anonymous." Again: "The encounter with the truth of art happens in the estranging language and images which makes perceptible, visible, and audible that which is no longer . . . perceived, said, and heard in everyday life" (quoted by Schoolman, 347). Schoolman concludes: "Bourgeois art [especially of the nineteenth-century romantics] is the only remaining sphere of criticism in a society wherein all forms of political and cultural discourse have bowed to the domination of technical reason" (347).

On one hand, this may seem high praise for traditional culture; it is the sole and unique source of revolutionary criticism of society left in a technological world. On the other hand—most obviously and most painfully—there is no one left today who can appreciate this critical potential of the great cultural moments of our past. At least, that is what Marcuse says.

My reaction to Marcuse's pessimism is complex. I agree that much that passes for art today lacks the critical potential of nineteenth-century art. Too much of it seems faddish and commercially motivated, aimed at achieving effect and success of the sort that will get it placed in contemporary art museums, or in the "new works" slot on symphony concert programs, or designed to "challenge" ordinary museum or concert audiences. Too little seems aimed at depicting grandeur and majesty or evoking profound and universal emotional responses.

At the same time, I also find myself disagreeing with Marcuse. It seems to me that both Bell and Marcuse are mesmerized by art critics, by mainstream museum and gallery art, and corresponding faddish music and dance and cultural filmmaking. These are what make "art news" media of various kinds in a commercialized world that favors a new model every year. But they are not all there is to the fine arts; much creativity in the arts never comes to the critics' attention. And besides, the fine arts are not by any means all there is to culture. Many ordinary citizens, at every level of the socioeconomic scale, find meaning in their lives and symbols to express that meaning. Devotees of so-called high culture typically look down upon manifestations of popular culture, but to most citizens today popular culture in all its forms helps give meaning to their lives.

There is, however, one area of culture that does seem to be in grave danger of extinction as a direct result of technological development. I have in mind the traditional cultures of Third World countries threatened by so-called technology transfer.

TRIBAL CULTURE AND MODERN TECHNOLOGY

I am not an expert on technology transfer to Third World countries, and my knowledge of the impact of technological development on traditional cultures in those countries is even more fragmentary, based as it is on nothing more than anecdotes related to me by students, through novels and movies, and in an occasional published account. I should also preface my brief remarks on this topic with a disclaimer: I do not like overromanticizing tribal cultures. That much said, I still believe that there are wonderful manifestations of the human spirit that once were to be found in tribal culture and that run the risk, as a result of technological development, of being turned into little more than curiosities in anthropological or craft museums.

Two recent accounts illustrate what I have in mind. Cheryl Bentsen in a recent book *Maasai Days*[11] depicts in a dramatic way the impact of Western development on one of the most romanticized of tribal cultures, that of the Masai. In a review in the *New York Times* (Sunday, 8 October 1989), Louisa Dawkins, a novelist-reviewer, captures the spirit of the book.

> Cheryl Bentsen deals with a very serious issue: how a people confronts massive social and economic change. . . . [She] undertook to discover how Masai men, women, teen-age boys and girls, caught between the traditional way that is no longer viable and the new way about which as yet not much is known, find their bearings.

Dawkins, following Bentsen closely, poses this pathetic version of the dilemma of meaning:

> Everybody talks about development, though nobody seems to know quite what it involves. Does it mean giving up the migratory herding life and settling down to cultivate maize and beans? Or becoming educated—and disaffected from your own people to the point that when you visit your mother, the dung and flies and mud of her *boma* [or encampment] disgust you and make you long for your neat concrete house in Nairobi?

Equally sad and even pathetic is a story I saw recently in the Toronto *Globe and Mail* about how tribal chieftains in Botswana, though they retain some of their traditional perquisites, are being systematically stripped of their powers in the name of "modernization." The story notes how the eight heads of the main tribal groups sit in a "house of chiefs" and are still allowed to exercise some traditional roles.

> At home in their villages, the chiefs exercise greater authority with fewer encumbrances than they do in the national capital. . . .
> Like several other African countries, Botswana has parallel legal structures that permit its tribal chiefs to rule [usually very fairly] on many legal matters. In seduction cases involving their daughters, for

example, Botswana men are likely to appeal to their chiefs rather than to a magistrate (*Globe and Mail,* 26 October 1989, A1–2).

The particular example might have been wasted on me, or might have seemed unimportant, if I had not seen, a short time earlier, an anthropological report that details the crucial significance of seduction and what we would call extramarital pregnancies in two tribal cultures, the Herero and Bushmen, in northern Botswana.

> The advantage to a woman of reproduction as a single woman is that she does not have to move away from home and live with a strange family—that of her new husband. . . . Daughters of poor families, on the other hand, provide a more certain inheritance for their children by marrying. A second advantage . . . may be that single women are able to retain closer control over and contact with their children.

In other words, seduction or marrying and similar customs are long standing and communally meaningful practices that the chiefs know how to deal with in a fairer manner than modern magistrate's courts. (The quote just cited is from a fascinating report in *Research/Penn State,* September 1989, on a study done among these tribes in Botswana by two Penn State anthropologists, Henry Harpending and Patricia Draper, that is part of a larger National Institute of Mental Health study of the meaning of aging in several cultures throughout the world. Harpending and Draper had been to Botswana twenty years earlier when the Bushmen still followed ancestral ways, before the government dug wells and settled them nearby.)

Whether the Masai or the Bushmen are better off or not—in terms of health or other alleged good effects of modernization—is not the issue. The point is that technological modernization often seriously affects the traditional sources of meaning for large numbers of people, often with no guidance on how to develop a new set of symbols to provide new meanings.

As my final source on the alleged negative influence of technology on culture, I want to turn to Jacques Ellul. His emphasis is not on culture generally, but on religion as an antidote to the technique that is crowding out spontaneity and value in all realms.[12]

ELLUL ON TECHNICIZATION AND THE POSSIBILITY OF RESISTANCE

Ellul is almost always referred to in American intellectual circles as espousing a single idea, that of autonomous technology. Interpreters sympathetic to Ellul say this is a distortion; some even go so far as to say that Ellul's seeming pessimism is really a warning. The situation he describes will inevitably get worse (they say Ellul means to say) unless we act quickly and decisively.

One recent interpreter of Ellul who says this is D. J. Wennemann:[13]

> It is my view that Ellul's thought has been misunderstood because of a lack of appreciation of the dialectical structure of his thought. . . . Reading *The Technological Society* in isolation, one can only conclude that Ellul is a technological pessimist of the most extreme sort. But this study must be balanced by contrasting it with its theological counterpart. Ellul has said that his plan of research was to oppose his study of technique with *The Ethics of Freedom*.

According to Wennemann, Christian revelation provides a possibility of absolute freedom, and he goes on:

> Ellul's intention is to attempt to make this freedom present to the technological world in which we live. In so doing, he hopes to introduce a breach in the technical system. It is Ellul's view that in this way alone [as Christians] are we able to live out our freedom in the deterministic technological world that we have created for ourselves.

Here is an extreme claim for the cultural value of religion; and, furthermore, it is possible to read the claim as calling for a return to traditional religion. However, it is also possible to see religion and other cultural manifestations as not reactionary but forward looking—as struggling to provide new meanings in troubled times.

DEWEY ON AESTHETICS AS PROVIDING MEANING FOR SOCIAL PROBLEM SOLVING

John Dewey is often interpreted as the great American advocate of social engineering; he sometimes is even parodied as an advocate of technocratic problem solving. Three new interpretations of Dewey seem much closer to what I think Dewey was saying. One places much greater stress than is customary on the aesthetic dimension in Dewey, making that the key to his (or any other) philosophy. The other two claim to move Dewey's view a step forward, turning pragmatism into an activist, even radical movement of *cultural criticism*.

The first of these reinterpretations of Dewey is that of Thomas

Alexander in *John Dewey's Theory of Art, Experience and Nature* (1987).[14] Alexander rejects interpretations of Dewey that focus exclusively or primarily on his instrumentalism. He asks: instrumentalism or social transformation for what? And he quotes Dewey as the authoritative interpreter of his own thought.

> To the aesthetic experience . . . the philosopher must go to understand what experience is. For this reason, . . . the theory of aesthetics put forward by a philospher . . . is a test of the capacity of the system he puts forth to grasp the nature of experience itself. (xiv, quoting Dewey's *Art as Experience*)

But Dewey and Alexander are here using both terms, "aesthetics" and "experience," in broad senses. Experience means the broad social experience of a community—what others would refer to as *popular culture* as well as *high culture*—and aesthetics includes not only the fine arts but also progressive religious attitudes and anything that creates meaning for a community.

> Experience [should be] understood from the Deweyan standpoint, as an involved, meaningful, and shared response to the world and to each other. . . . To keep experience from being taken in a subjective or reductionistic manner, one requires a theory which will maintain its situational and transactional features in full view. To keep experience from being treated always as a form of cognition . . . , one needs to articulate a position where the larger issues of human meaning and value contextualize the pursuit of knowledge. Knowledge is only possible because we can respond to the world as a dramatically enacted project in which meanings and values can be won, lost, and shared.

On the breadth of experience (as focused by its most significant achievements):

> Works of art are significant for Dewey. While an inquiry into the origin and nature of the aesthetic can profitably begin by forgetting examples of fine art and focusing on the implicit aesthetic possibilities of our daily living . . . , such an inquiry must eventually recall and reflect upon truly great works of art. These reveal as almost nothing else can . . . the inherent possibilities of experience. (250)

And on religion:

> This aspect [of social action being carried out within a broad and undefined cultural background] is precisely what Dewey describes in *A Common Faith* as the religious *quality* which experience may have.

Dewey thereby is able to distinguish "the religious," which is a non-cognitive quality, from the particular doctrinal beliefs which constitute religions. (255)

This is what allows Dewey to praise religion as a meaning-creating accompaniment, indeed the goal, of social activism at the same time that he so often savagely attacks traditional religious denominations as enemies of progress.

To summarize, according to Alexander—and I think he is unquestionably correct here—the progressive social activism that Dewey espouses in the name of "scientific method" or "instrumentalism" can only be progressive if it aims at creating meaning for a particular community struggling to achieve meaning in troubled or problematic times.

A second recent interpretation of Dewey that I find helpful is Larry Hickman's *John Dewey's Pragmatic Technology* (1990).[15] Building on Ralph Sleeper's *The Necessity of Pragmatism: John Dewey's Conception of Philosophy* (1986)[16] and Sleeper's thesis that Dewey's philosophy is primarily an instrument for cultural transformation, Hickman goes a step farther in maintaining that Dewey's philosophy is explicitly and consciously a meliorist critique of *technological* culture. Like Alexander, Hickman discusses Dewey's aesthetics, focusing in particular on the lack of a clear distinction, in Dewey's thought, between the fine arts and the ordinary workaday life of ordinary citizens in our technological culture. Hickman also details the social and political ramifications of Dewey's critique of technological culture.

A third interpretation of Dewey is even clearer about the political implications of Dewey's critique. That is Cornel West's *The American Evasion of Philosophy: A Genealogy of Pragmatism* (1989).[17] West claims to be improving upon, even going beyond Dewey to a "prophetic pragmatism" that learns more from Marxism than Dewey was willing to do. But I think West is more correct when he also claims that his view is a culmination of a pragmatic tradition in American thought that has Dewey as its pivotal figure, although for historical reasons it reaches its culmination today in the intellectual and cultural environment of the U.S.

West would agree with Alexander's interpretation of Dewey to the effect that "knowledge is only possible because we can respond to the world as a dramatically enacted project in which meanings and values can be won, lost, and shared" (as quoted earlier). But West is much more precise in pointing out that this requires historical consciousness of particular historical movements with their struggles for new cultural meanings against entrenched traditions.

The particular cultural context of the U.S. struggle today for meaning in a technological world that West concentrates on is the conflict in the universities over including in the canon of meaningful discourse new voices arising from black experience, women's experience, Hispanic experience, in short, of so-called minority experience generally. It happens that this struggle for the inclusion of multicultural diversity coincides precisely with the movement called *postmodernism,* a loose term found in cultural studies of all sorts, from critiques of art and architecture to philosophy, but also a term that is most often associated with the jaded neoconservatism of authors such as Richard Rorty.[18] (Rorty is West's principal target in his book.) West says that we in the U.S. today need a repudiation of this postmodern pseudoculture; and that we need to fight against it, faithful to Dewey's impulse, in a power struggle to broaden democracy and include the aspirations of those peoples West lumps under Frantz Fanon's label "the wretched of the earth," both in U.S. culture itself and in the Third World.

People say that ours is a technological world; some even speak of ours as a technological culture. Daniel Bell and Herbert Marcuse and Jacques Ellul would say it is a technological anticulture. Dewey, as I interpret him, might well admit that technology today threatens traditional culture—and perhaps nowhere more than in the Third World. But Dewey would take this not as grounds for pessimism but as a challenge to social activism; as an invitation to create new symbols and new meanings for our troubled and problem ridden world.

It is at this point that I would raise a question: Are there activists in sufficient numbers to carry on the struggle that West calls for? Is anyone really fighting to establish new meanings for our technological world? West himself believes that there are, and he points to the feminists, blacks, Hispanics, and other minorities who have been leading the battle for multicultural diversity in American universities. But what he really means to say is that the constituencies for which these academics speak—the real wretched of the earth—need to struggle with the cultural powers that be in order to create a new and more meaningful culture in a world threatened on every side by forces labeled "technological progress" that are in fact rarely progressive for them. And when stated in such polarized terms, it is not clear that technological nihilism will not triumph over a new, progressive, more open, technological democracy aimed at including the poor and the oppressed. This, however, would not have dismayed Dewey any more than it does me or West; a meaningful existence is not something to be taken for granted, but something that must be won through arduous social struggle.

NOTES

1. Daniel Bell, *The End of Ideology: On the Exhaustion of Political Ideas in the Fifties; With a New Afterword* (1962; reprint, Cambridge: Harvard University Press, 1988).

2. Daniel Bell, *The Coming of Post-Industrial Society: A Venture in Social Forecasting* (New York: Basic Books, 1973).

3. Daniel Bell, *The Cultural Contradictions of Capitalism* (1976; reprint, New York: Basic Books, 1978).

4. Friedrich Nietzsche, *The Will To Power* (New York: Random House, 1967).

5. Joseph Conrad, *The Secret Agent* (1907; reprint, Garden City, N.Y.: Doubleday, 1953).

6. Emile Durkheim, *The Rules of Sociological Method* (Glencoe, Ill.: Free Press, 1938).

7. Talcott Parsons, *Essays in Sociological Theory, Pure and Applied* (Glencoe, Ill.: Free Press, 1949).

8. John Kenneth Galbraith, *The New Industrial State* (Boston: Houghton Mifflin, 1967), and *Economics and the Public Purpose* (Boston: Houghton Mifflin, 1973).

9. Herbert Marcuse, *One-Dimensional Man* (Boston: Beacon, 1964) *Counterrevolution and Revolt* (Boston: Beacon, 1972), and *The Aesthetic Dimension* (Boston: Beacon, 1978).

10. Morton Schoolman, *The Imaginary Witness: The Critical Theory of Herbert Marcuse* (New York: Free Press, 1980).

11. Cheryl Bentsen, *Maasai Days* (New York: Summit, 1989).

12. Jacques Ellul, *The Technological Society* (New York: Knopf, 1964), and *The Ethics of Freedom* (Grand Rapids, Mich.: Eerdmans, 1976); see Joyce M. Hanks, compiler, *Jacques Ellul: A Comprehensive Bibliography* (Greenwich, Conn.: JAI Press, 1984), for a virtually complete list of works by and about Ellul.

13. D. J. Wennemann, "An Interpretation of Jacques Ellul's Dialectical Method," *Broad and Narrow Interpretations of Philosophy of Technology,* ed. P. Durbin (Dordrecht: Kluwer, 1990), 188.

14. Thomas A. Alexander, *John Dewey's Theory of Art, Experience and Nature* (Albany: State University of New York Press, 1987).

15. Larry Hickman, *John Dewey's Pragmatic Technology* (Bloomington: Indiana University Press, 1990).

16. Ralph W. Sleeper, *The Necessity of Pragmatism: John Dewey's Conception of Philosophy* (New Haven: Yale University Press, 1986).

17. Cornel West, *The American Evasion of Philosophy: A Genealogy of Pragmatism* (Madison: University of Wisconsin Press, 1989).

18. Richard Rorty, *Philosophy and the Mirror of Nature* (Princeton: Princeton University Press, 1979), *Consequences of Pragmatism* (Minneapolis: University of Minnesota Press, 1982), and *Contingency, Irony, and Solidarity* (New York: Cambridge University Press, 1989).

New Technologies and an Old Debate: Implications for Latin America

MARGARITA M. PEÑA

We want development . . . with siesta
—Jorge Sábato, Argentinian scientist

NEW TECHNOLOGIES IN THE DEBATE ON DEVELOPMENT: A WORKING HYPOTHESIS

My purpose in this article is to present the most representative positions in the debate on science and technology in the Third World, to note their implications for Latin America, and to criticize certain aspects of each of these positions. My thesis is that the development debate has been dominated by two main themes: the absence of local science and technology; and the integration of modern science and technology into economic development policies. Essentially, my argument is that current analyses of scientific and technological issues are critical, and sometimes radical, regarding all factors but one: the technologies themselves and their scientific base. All sides share a blind faith in the potential for personal and social progress of the Western scientific and technological tradition. Though few would argue that the Latin American countries are on the periphery of contemporary scientific and technological knowledge, and that this marginality is to a large extent responsible for the stagnation of their economies and of the profound economic problems they currently face, Latin Americans ought to be careful regarding the promises of new technological advances. The absence of a critical attitude precisely toward technical knowledge may lead us to a situation in which the remedy is worse than the illness itself.

This became a concern of mine some years ago, while I was conducting research on the relations between education and technological development in Colombia. My work put me in direct contact with government documents and with some of the people who were, at that time, directly involved in the formulation of a scientific and technological policy for a country undergoing severe economic and social problems. The government was faced, then as it is now, with the imperative of implementing policies aimed at the elimination of the causes of a deep social unrest, of which violence was, and still is, the most eloquent expression: unemployment, homelessness, ignorance, and widespread poverty. Some of those problems were explained as being the result of the backwardness of the production sector of the economy, the obsolescence of the education system, or, in sum, as the consequence of the lack of proper mechanisms capable of providing a solid foundation for economic and social policies. It is not surprising that the technologies, with all their promises, became an important component of the development strategy. I was infected with that optimism, and the conclusions of my research pointed in the same direction.

Although it was not so long ago that I wrote those conclusions, it is now possible for me to give a different interpretation to the information I collected, from which I take most of the examples for this article. While these examples, drawn from Colombia, are not enough to generalize to all of Latin America, let alone to all of the Third World, they do offer starting points for the identification of related trends in other countries. They should help us gain a better understanding of the problem posed by modern technologies, and technical knowledge, from a Latin American perspective, thus overcoming the restricted nationalistic framework that has dominated the technological development rhetoric.

The progressive social implications of new technologies is one of the most important characteristics of the current model of development in Latin America. During the last thirty years, a belief in industrialization as the carrier of a desirable new economic and social order has dominated debate on the way to achieve an appropriate level of "development" and to overcome "backwardness." According to this thesis, the differences between rich and poor countries come from the monopoly by the former of industrial production, and more recently, of access to the scientific and technical knowledge upon which industrial production is based.

What some view as the failure of this economic model—increasing poverty for many, Latin America's subordinate position in the world economic system, its debt crisis—has begun to give shape to a marked skepticism regarding Latin America's capacity to generate development and its contribution to social welfare. This reaction has been important in some academic circles, in which the expression *counterdevelopment* has been introduced into the debate. Proponents of this view suggest that the terms of the debate do not correspond to a desirable end, but rather to an ideology that legitimizes certain realities and certain policies. This ideology has ended up shaping the world we know: a world in which a privileged minority is more "advanced" than the majority. At the same time, this ideology has provided the setting for the formulation of a value system according to which being more developed is "better," and all that is done to overcome "backwardness" is morally justifiable. The diffusion of Western technologies and the destruction of cultures that have accompanied the modernization of the non-European world has been justified in the name of progress, the most powerful concept of this ideology.

I do not intend, at least for the moment, to elaborate arguments for or against such views. I am interested, instead, in using them as a starting point, because they question the validity of the values that have accompanied the modernization process as experienced in Latin America. Among these values are efficiency and the subordination of nature to the needs of economy, which in modern history have been raised above all other values, particularly those of justice and respect for the environment. At best, this value system is the result of a naive optimism, according to which progress would be the basis of well-being and happiness for all.

A critique of technology that integrates examination of this value dimension into a discussion of technology has not been accomplished in Latin America. From inside the social system, the *dependentistas,* warn that First World technology imports contribute to reinforcing the peripheral condition of Latin American economies. But they do not focus their criticism on technology. Instead, they focus on the mechanisms of technology transfer and diffusion, and on the absence of a local science capable of sponsoring indigenous technological development. A more adequate critique of technology per se might have been provided by the followers of the so-called appropriate technologies movement. This movement, however, has focused more on making, that is, on producing (clever) artifacts, than on formulating a proposal for technological development that better conforms to the social needs of the region.[1]

Against the background of these critical positions, new technologies came into existence during the last decade. There was talk of a second, or even a third, industrial revolution. The new technologies promised more flexible forms of production, unlimited access to information, more and better agricultural production, less illness, new materials with extraordinary qualities—in short, a new age of science and technology, and with it a new wave of technological euphoria. The criticism of technology, never solid enough in Latin America, was eclipsed by the image of a better world based upon the new technologies. To oppose them was the same as opposing progress, that is, to oppose the most extraordinary achievements of human intelligence.

In the event, the diffusion of new technologies seems actually to have exacerbated the backwardness syndrome in Latin American societies and to have produced a collective anxiety about the need to catch up. The education system is blamed for deficiencies in scientific and technological education. The harshest critics make an effort to distribute responsibilities, blaming industry for not offering attractive opportunities to local professionals, and the government for not promoting an economic and industrial policy that serves as an incentive for scientists and technologists to stay in their own countries. From the political point of view, the popularity of "technological development programs" and "democratization campaigns" is guaranteed, suggesting populist optimism concerning technology.

This optimism may sound strange at a time when the environmental crisis, itself a product of life-styles keyed to the new technological developments that Latin American societies were trying to imitate, had acquired a global character, overlapping national and regional frontiers. Nevertheless, the technological development debate is almost always presented as a national debate, inspired by the ideal of political and economic self-determination. One cannot, however, label it entirely as "populist" or "democratic." Since its insertion into the world economic system, about five centuries ago, Latin America has responded, or has had to respond, to demands that have been imposed on the region by a Europe based sociotechnical system: demands first for metals, later for agricultural products, energy resources, cheap labor, and latterly, as markets for industrial products. Though these demands are claimed to be based on the role that each country has in an almost natural international division of labor, history has shown that the supposed interdependence that the system trumpets is nothing but a disguised relationship of unequals, separated from each other by an ever-widening gap. Thus, it is understandable that development projects of any establishment political party consider technology, and more recently new technologies, as the cornerstone of any program of economic development.

THE ABSENCE OF A LOCAL TECHNOLOGICAL AND SCIENTIFIC CAPACITY: VARIATIONS ON THE DEPENDENCY THEME

In general, those interested in Latin American issues, regardless of what point on the political spectrum they occupy, point to the absence of local scientific and technological capabilities as one of the factors that has contributed most to perpetuating the peripheral condition of these countries in the world economic system, and to aggravating their social problems. The subordination of local economic activity to technologies and to knowledge developed in advanced countries puts the receiving countries at a disadvantage, because both technologies and knowledge are commodities that poor countries must purchase in the world market, investing in them financial resources that could be put to better use in social development projects. These countries are compelled to acquire such goods because of their need to renovate their production systems. This renovation is, in turn an indispensable condition for keeping competitive in the world market, thus guaranteeing access to the foreign exchange they need to purchase the goods and services (technology included) that their peripheral condition does not let them generate by themselves.

In the context of dependency theory, technology is one of the mechanisms through which the subordination of the periphery to the center is perpetuated. In its classical formulation, technological dependence was explained as the result of the absence, in peripheral societies, of a capital goods sector to their economies, the sector in which, until very recently, the main technological innovations originated. Given their lack of this "internal core of creativity," and their inability to develop it because of their peripheral condition, the Third World countries were doomed to remain "underdeveloped," or, in other versions of the theory, to achieve only a "dependent development."[2]

This thesis was invalidated by studies made in several Latin American countries, as well as in India and South Korea.[3] These studies showed that it was possible for less developed countries to master foreign technology. For example, as a result of the process known as "technological learning that happens by doing" on the shop floor, there have been developed innovations so important that some countries have been able profitably to export modified products and processes of their own. On the other hand, with the so-called scientific-technological revolution, that internal core of creativity has been displaced from the shop floor to the laboratory, and a clear distinction between innovation and invention has been established. Innovation happens from the bottom up as a result of interactions between people and machines or processes; invention is the result of targeted research aimed at the development of new products and processes and requires a solid scientific base.

The studies to which I have referred reveal important characteristics of the technological development process in some Latin American countries, but they do not render the technological dependence thesis invalid. On the contrary, it is easy to demonstrate that even countries that have the capacity to export innovative technology have not entirely cast off their dependency bonds. In search of a more successful explanation, another variable has been identified as the direct cause of dependency, namely, the marginality of science with regard to production.[4] These analyses emphasize the outward orientation of the scientific communities in the Third World, which are, in most cases, made up of people trained in the kinds of science done in the advanced countries. These people are interested mainly in topics that have been made legitimate by the international scientific community, showing no concern for regional development needs. This explains why pure science publications originate in Third World research institutions that were initially established with the purpose of doing research directly related to sector development policies. A study of the links between research and diffusion of technology, carried out by the Colombian Institute for Agricultural Research (ICA, for its Spanish initials), revealed that the *Revista ICA,* the institute's most important scientific journal, does not have much value for ICA extension workers, "due to its sophisticated format, because *its objective is to show the national and international community the latest advances made by ICA in the agricultural sciences*" (emphasis added).[5] The seriousness of such a statement becomes clear when it is compared to the objectives that are supposedly pursued by the institute: "to solve the technical problems of commercial and subsistence agriculture, to raise the food producing capacity, to improve the standard of living of the rural peasantry, and to contribute to the generation of foreign exchange."[6]

Thus, the problem is not that science is not being done. There is scientific activity, but of a dependent kind. Its local legitimacy, based on regional needs, has become subordinated to its being accepted by the international scientific and financial agencies.[7] Moreover, it is done under such restricted financial conditions, that that alone could be the reason why so little of significance is accomplished.

This explanation of technological dependence exposes the structural limitations to the emergence of local versions of modern science and technology in the Third World, and it sets the topic in the context of center-periphery relations. The argument thus follows the same logic as the one that explains technological dependency as the result of the lack of a capital goods sector in the sense that it explains dependence in terms of the absence of an internal core of creativity. It also insists on the need of creating one, or strengthening it, in this context in the form of the applied science research center, in a way that responds to the needs of the process of development.

It helps to point out that those who have interpreted the problem in terms of the absence of local science have referred mostly to a lack of basic science, which was expected to tie in with development policies through the classical chain of basic research—applied research—technological development—dissemination. This is the model followed by institutions such as ICA, even in the face of known results regarding the divergence between basic research and the innovations that are actually disseminated. Also, this view tends to dominate among members of the Latin American scientific community who show concern for social and economic problems.

However, as the experience of the industrialized countries seems to demonstrate, new technologies do not come from basic research alone, but from research and development (R and D) activities. The R and D laboratory is now the internal core of creativity whose absence explains why some countries create new technologies and others have to acquire them in the market. According to Martin Fransman,[8] technological capabilities are of two kinds: "know-how," or the capacity to search, select, master, adapt, and innovate foreign technology; and "know-why," or the capacity required to develop new products and processes (or what is known as research and development), and to undertake basic research. Research and development (R and D) differs from basic research, because it looks for concrete and economically feasible results.

Research and development activities are virtually nonexistent in the Third World. Latin American countries invest ten times less of their gross national product in R and D than does any industrialized country; the United States alone spends more money in R and D in a year than do all Latin American countries combined in most decades.[9] To make things worse, advances in "know-how," or in the capacity of a country to innovate, do not automatically lead to the acquisition of a "know-why." For this evolution to occur, a qualitative jump that includes a change in the appropriate political and economic conditions, has to take place: the creation and endowment of R and D laboratories, the education of the highly skilled human resources needed, the creation of a national market for research results, and so on. In the case of South Korea, for example, even with the rapid advancements in know-how, industries did not first master the engineering knowledge supporting such progress, with the result that the country's capacity to operate factories has been superior to its capacity to design them. Recent efforts by government and industry in South Korea are directed toward the establishment of R and D laboratories with the aim, in turn, of nurturing that inventive capacity that makes possible the maintenance of a country's competitive position in the world market.

It could thus be said that the absence of know-why is the main characteristic, if not the cause, of the technological dependence that afflicts even the most developed economies of the Third World. In spite of the existence of high technology enclaves on the periphery, the restructuring of the world economy that has been made possible by the scientific-technological revolution has resulted in an ever-widening gap separating center and periphery.[10] The high technology comes sealed in "black boxes," and it depends on the provider for its innovation, maintenance, and reproduction. The periphery is thus at the receiving end of a process, each stage of which is built upon the preceding one. The transference of R and D happens mostly in low technology sectors, which are characterized by a low level of monopolization. Thus, the division of labor in R and D is even stricter than the one in production.[11]

It would thus seem possible to conclude that, at least in theory, the existence of R and D activities in the periphery would reduce technological dependence. It would make possible the local production of products and processes that must otherwise be bought, at high prices, in the international market. In the case of biotechnology, for example, the lack of R and D will lead to greater dependence, which, if maintained, will keep these countries at the margin of the international food market, with a consequent weakness of national economies.[12]

The main limitation of a strictly *dependentista* reading of science and technology issues in the less developed countries seems obvious: overcoming the dependent condition does not guarantee the solution of the economic and social problems that gave rise to this kind of criticism in the first place. By focusing too much on the need to strengthen the position of Latin American countries in the international market, this view offers too simplistic an interpretation of the problem, in which development is but a side effect, if not a moral justification, of policies aimed at strengthening the external sector of the economy under the assumption that this will bring ample benefits to the great masses of the population. The falsehood of such an assumption has been demonstrated in the case of Brazil, a country that has made important advances in constructing a local capacity in science and technology, in the process solving a number of pressing problems. Unfortunately, they have not been able to solve others. Great problems still plague Brazilian society, particularly an astronomical foreign debt that has brought the economy to its knees in order to meet the demands of international bankers, if Brazil is to attract the foreign exchange it needs to fulfill its financial obligations.[13]

A second limitation of the *dependentista* interpretation is also obvious, but to make it explicit is particularly important under the current circumstances. By focusing exclusively on the forms of scientific and technological knowledge upon which the superiority of the Western tradition is based, this interpretation completely ignores the connection between that tradition and the growing deterioration of nature. The deterioration of the environment is not only a consequence of the rapacity of capital, it is also an effect of our consumption habits and of the technologies used to satisfy those habits. Awareness of this important dimension is not reflected in the writings of the majority of Latin Americans who have viewed the development problem from the perspective of political economy.[14] In the absence of such a dimension, the *dependentista* interpretation of the scientific and technological problem in Latin America leaves us without any tools to undertake the task of thinking seriously about a desirable scientific and technological future for the region, which is, ultimately, the kind of reflection that should interest us most.

SCIENCE AND TECHNOLOGY FOR DEVELOPMENT

It is appropriate to ask whether more recent formulations of the problem have overcome the limitations of the *dependentist* approach. Taking a slightly different perspective, many analysts have insisted that real development is generated internally, with local economic activity that is strong enough to generate employment and expand the internal market, and not externally, in the form of foreign investment, economic openness, and production of export goods, as established under development plans imposed upon poor countries by international leaders. This "progressive" sector advocates the creation of a local scientific-technological capacity in the periphery, but it has also pointed out that countries that have not been able to satisfy the basic needs of the population must develop the kind of science and technology that can effectively contribute to social well-being.

In other words, scientific and technological advances must be geared to the solution of the basic needs of the population, such as housing, clothing, food, and health. If, as we have learned from painful experience, technologies developed for countries that have already met those needs are not appropriate for our countries (as has been shown, for example, by several experiments in the field of nutrition), the selective adoption and development of technologies that can be "made suitable" to those needs would, nonetheless, effectively contribute to real social progress. This approach has been spelled out in some national science and technology plans. For example, the Colombian 1987–90 science and technology plan proposed to integrate science and technology within "policies aimed at reducing poverty and generating more employment," and to "selectively connect *the most advanced science and technology* to the production of goods and services, in search of a better quality of life" (emphasis added).[15]

These "appropriate forms" do not exclude the newest and most sophisticated technologies. On the contrary, there seems to be a great deal of optimism about the advantages that new technologies offer for the achievement of balanced development.

Microelectronics and the new biotechnologies are those around which the greatest expectations have been raised. It has been suggested, for example, that information technology will allow the establishment of small and flexible manufacturing firms geared to the internal market and not for export.[16] This is so, because the technology of computerized numerical machine tool control makes it possible for small batch production to be as profitable as has been specialized machinery for mass production of homogeneous products.[17]

Beside the industrial applications of microelectronics, applications in other fields have awakened enthusiasm: the modernization of public services and administration, the "informatization"[18] of trade and banking, access to data bases, and advances in telecommunications. One could add to this an optimism about the educational applications of information technology: teaching machines that interact with students, and make learning dynamic, thereby contributing to an improved quality of education.

Little has been written about the cultural effects of the indiscriminate dissemination of information technologies in societies that have not yet completed the modernization process. However, in the case of microelectronics, it is possible to find some warning about negative impacts, particularly unemployment and skill obsolescence. This is not the case with respect to the new biotechnologies, which have raised such high expectations. Here the potential negative impacts of new biotechnologies are not sufficiently known. The discussion tends to focus more on the promises, the new vaccines, biological pest controls, and enhanced food production,[19] rather than on the problems they might cause. At most, proposals in favor of encouraging biotechnology research in Latin America face two kinds of criticism. One has to do with the absence of a solid scientific base, one capable of supporting efforts to develop a national biotechnology. The other refers to the dangers that the connection between genetic engineering and corporate interests in the industrialized countries represents for the Latin American economies. In contrast, the hazards to which we may be exposed, if we accept any given advance in genetic engineering, have not been sufficiently researched. This explains why countries that, like Colombia, are heavily dependent on the use of pesticides to maintain the yield of agricultural production see the development of biological pest controls, or microorganisms that fight the dangerous species, as viable alternatives to offset the indiscriminate use of chemical pesticides.[20] But they do so without much regard for the possible disastrous consequences that the propagation of microorganisms may have on the environment.[21]

From a progressive standpoint, these and other objections in response to the new technologies may be avoided with a healthy scientific-technological policy. Combined with social and economic policies, science and technology, can make it possible for Latin American economies, traditionally externally oriented, to implement strategies aimed at the creation of industries and the expansion of agriculture, the creation of jobs, the strengthening of the internal market, and the progressive reduction of dependency in relation to the world economic system.

In this effort to integrate adequately science and technology with demands for development, R and D activities must play a central and constructive role. As Stephen Hill has mentioned, Third World research in science and technology must fill the gaps left by research done in the industrialized countries. This will be possible, if research projects are formulated around problems that need solutions, rather than around problems promising only in terms of their legitimacy in the eyes of the international scientific community.[22] This requires scientists and technologists to be well prepared in their fields of expertise, but also knowledgeable about the problems that should be addressed by their research. Thus, another dimension should be added to the know-why: the "know-what-for" or the awareness of what and who will be served by the research results.

It goes without saying that the presence of social objectives in national science and technology policies does not necessarily express an effective commitment on behalf of the governments. In fact, it is quite frequently the case that specific economic policies are in conflict with more general ideas about job creation and production for the domestic market. In many cases, documents of this sort merely fulfill the political function of satisfying certain sectors of the public and of demonstrating that the government wishes to solve the problems of society. Therefore, it is not surprising that progressive discourse survives in conditions otherwise so adverse.

Another reason why such a discourse is tolerated is an awareness existing within the Latin American political establishment that investment in science and technology is, in the long run, a good thing. Latin American leaders have been educated in the most classical European tradition, and they have internalized concepts and techniques of the empirical social sciences at the top of which are efficiency and its measurement. They know that science is useful and necessary to solve problems. It is quite another thing, however, to have the will, or the power, to make an effective commitment to this ideal.[23]

The question I want to raise, finally, is whether, besides expressing good intentions, progressive science and technology policies can indeed contribute to a change in the economic conditions in Latin America and to a qualitative improvement of the living conditions of the population. In other words, can, under appropriate political conditions, the selective adoption and/or development of new technologies contribute to the fulfillment of the social ends of economic development? Ultimately, the question is whether "development with siesta," as referred to by Sábato—"development Latin American style"—is at all possible. That is, can we have all the advantages of development, without having to give up the customs, cultural values, and ways of life so profoundly rooted in tradition and associated with quality of life that has already been lost, maybe forever, in the most advanced countries?

This proposal is attractive but still very problematic. Even if we hold that development is a desirable end—an arguable thesis, as I suggested in the beginning of this essay—or that development is desirable provided that it is no longer defined as modernization but as emancipation—an interesting idea well worth exploring—science and technology can hardly be considered means by which to achieve one result only. An attitude of this kind remains totally uncritical regarding the technologies involved. At best, it is concerned wtih the positive or negative impacts they may have on the society and the environment. In general, such an attitude tends to overlook the fact that, while there are technologies whose contribution to the improvement of the quality of life is obvious, there are also technologies that are specifically destructive. The technologies of torture, and of defoliation, so widespread in Latin America, are eloquent examples of what the Costa Rican philosopher Edgar Roy Ramírez has called "pernicious technology."[24]

This particular way of looking at technology as something external to the social process, over whose impacts, good or bad, some kind of control may be exerted, places technology on a neutral ground and obscures the fact that technologies are, as Langdon Winner has put it, "forms of life," in the sense that most of the transformations that are produced by technological innovation are really variations on ancient patterns. Far from being mere impacts or side effects, these transformations constitute the most important aspect of any new technology.[25]

In the case of Latin America, a critique of technology that incorporates the concept of technologies as forms of life has yet to be developed. An appropriate analysis has to go beyond—if we accept Winner's view—a mere identification of the technologies we want to own and at what price, to a reflection upon the kind of world we will be making when we put them in place. In Winner's words, "This suggests that we [need to] pay attention not only to the making of physical instruments and processes, but also to the production of psychological, social, and political conditions as a part of any significant technological change."[26] The complete analysis and appraisal of specific technologies thus becomes an unavoidable task. If we do not do it, we may achieve development, but we may also have to give up the siesta.

NOTES

1. A critical reading of the evolution of the movement toward "appropriate technologies" has yet to be done. Its marginality in the debate on development may be due to the fact that its proponents fell short in the formulation of a global and realistic proposal directed toward demonstrating what can be done in the field of "appropriate" artifacts and processes, which in turn can also be integrated into broader social development projects and be used in the strengthening of local

economies. With some exceptions, those efforts were not part, nor were config-ured as, political projects: they were put forth as alternatives that could subsist within the established system. Ultimately they were asphyxiated by that system.

2. Fernando Henrique Cardozo, "Dependency and Development in Latin America," in *Introduction to the Sociology of "Developing Societies,"* ed. H. Alavi and T. Shanin (London: Macmillan, 1982), 120; Moisés Ikonikoff, "La industrialización del mundo a prueba de los grandes cambios," *El Trimestre Económico* 54 (1987).

3. Jorge Katz, ed., *Technology Generation in Latin American Manufacturing Industries* (New York: St. Martin's Press, 1987). For the Indian case see Sanjaya Lall, "Technological Learning in the Third World: Some Implications of Tech-nology Exports," in *The Economics of New Technologies in Developing Coun-tries,* ed. F. Stewart and J. James (London: Frances Printer, 1982). The Korean case is analyzed by Larry E. Westphal, Yung W. Rhee, and Garry Pursell, in "Sources of Technological Capability in South Korea," in *Technological Ca-pability in the Third World,* ed. M. Fransman and K. King (New York: St. Martin's Press, 1984).

4. C. Cooper, "Science, Technology and Production in the Underdeveloped Countries," *Journal of Development Studies* 5, no. 2; Amilcar Herrara, *Ciencia y Política* (México, D.F.: Siglo XXI editores, 1981); Frances Stewart, *Technology and Development* (London: Macmillan, 1977).

5. Germán Urrego, *La articulación entre la investigación y la transferencia de tecnología. El caso del fríjol en el sur del Huila* (Bogota: ICA-ISNAR, 1988), 68–69.

6. Instituto Colombiano Agropecuario, "Algunas áreas de investigación del ICA y su articulación con el sector productivo" (Paper presented in the Encuentro Nacional sobre Desarrollo Tecnológico Industrial, Medellín, Colombia, October, 1985).

7. In the words of an Argentinian scientist, "The topics of our research are determined in the research centers of the north, the techniques are invented there, the instruments are designed and generally constructed there, and they have even taken our language away, due to the fact that the journals we read are in English. The mark of achievement is to contribute to the development of science in the European or North American centers, and it is increasingly frequent to make pilgrimages to those centers in order to legitimize our research." Otto T. Solbrig, "¿Destrucción o transformación del paisaje tropical sudamenricano?" *Intercien-cia* 13, no. 2 (1988): 79.

8. Martin Fransman, *Technology and Economic Development* (Boulder: West-view Press, 1986).

9. Aaron Segal, "Latin America. Development with Siesta," in *Learning by Doing: Science and Technology in the Developing World,* ed. A. Segal (Boulder: Westview Press, 1987), 38.

10. I refer to the fragmentation of the productive process, which makes possible its localization in various parts of the globe. This is a characteristic of what is known as the New International Division of Labor.

11. H. Hveen, "Selective Dissociation in the Technology Sector," in *The Anti-nomies of Interdependence: National Welfare and International Division of Labor,* ed. J. G. Ruggie (New York: Columbia University Press, 1983), 283–84.

12. "Hacia una biotecnología del Tercer Mundo," *Interciencia* 14, no. 6 (1989): 350.

13. Amiya Kuyar Bagchi, "Technological Self-reliance, Dependence and Un-

derdevelopment," in *Science, Technology and Development,* ed. A. Wad (Boulder, Westview Press, 1988), 83.

14. Michael Redclift, *Development and the Environmental Crisis: Red or Green Alternatives?* (London: Methuen, 1984), 17.

15. COLCIENCIAS-MEN, *Plan Nacional de Ciencia y Tecnología para una Economía Social* (Bogotá: COLCIENCIAS), 17.

16. See, for example, Pedro Amaya, "Bases para una política nacional en Ciencia y Tecnología," in *Foro Nacional sobre Política de Ciencia y Tecnología para el Desarrollo. Memorias* (Bogotá: COLCIENCIAS, 1987); S. Cole, "The Impact of Information Technology," *World Development* 14, no. 10/11 (1986); Atul Wad, "Microelectronics: Implications and Strategies for the Third World," *Third World Quarterly* 4, no. 2 (1982).

17. Daniel Chudnovsky, "The Diffusion and Production of New Technologies: The Case for Numerically Controlled Machine Tools," in *Science, Technology and Development,* ed. A. Wad, (Boulder: Westview Press, 1988), 172.

18. This is a literal translation of the Spanish *informatización,* a word used widely in the Spanish speaking world to refer to the introduction of information technology in human activities.

19. See, for example, Albert Sasson, *Biotechnologies and Development* (Paris: Unesco, 1988); R. A. Zilinskas, "The International Centre for Genetic Engineering and Biotechnology. A New International Scientific Organization," *Technology in Society* 9 (1987): 47–61.

20. "El cultivo de la muerte," *Interciencia* 12, no. 5 (1987): 250.

21. As José Sanmartín has pointed out, the impact of technologies, intended not only to control and manipulate natural causes, but also to replace natural products (the difference between a seed improved by germinal treatment and organisms created by recombinant DNA techniques), is the uncontrolled addition to nature of materials and organisms whose dissemination and survival we may not be able to control. José Sanmartín, "No toda producción es síntesis," *Anthropos,* no. 94/95 (1989): 41.

22. Stephen Hill, "Basic Design Principles for National Research in Developing Countries," *Technology in Society* 9 (1987): 67–68.

23. In the opening speech of the sixth meeting of the Andean Parliament, Virgilio Barco Vargas, then president of Colombia, expressed that "science and technology must occupy their proper place, as engines of progress in an era marked by the innovation of knowledge and its immediate repercussion in the progress of the countries and the well being of the people. The quality of life depends directly as it has recently been verified by important economic analysts, on the investment that the state makes in the people, in the improvement of their intellectual potential and of their capacity for analysis and innovation." In COL-CIENCIAS-MEN, *Plan de Ciencia y Tecnología para una Economía Social* (Bogotá: COLCIENCIAS), 17.

24. Edgar Roy Ramírez, "El argumento tecnológico, la tecnología perniciosa y la ética" (Paper presented in the I Interamerican Congress of Philosophy and Technology, Mayagüez, Puerto Rico, October 1988).

25. Langdon Winner, *The Whale and the Reactor: A Search for Limits in an Age of High Technology* (Chicago: University of Chicago Press, 1986), chap. 1.

26. Ibid., 17.

The Nature of Global Processes

RICHARD WORTHINGTON

INTRODUCTION

For most of human history, people have viewed themselves as being fundamentally connected with society, nature, and the cosmos. The reconceptualization of nature as a machine at the dawn of the modern era tore these bonds asunder, clearing the path for modern science, individualism, and a seemingly boundless expansion of individual mobility, material wealth, and human progress.

The flip side to this story is by now no secret, and the approaching quincentenary of Columbus's voyage to the New World marks an appropriate occasion for reevaluating the expansion of the West and the prospects for a new kind of growth. The consequences of the West's intellectual and practical fragmentation are manifest not only in environmental breakdown, but also in ruinous political and military competition, widespread addiction to powerful drugs, and myriad other well-publicized maladies. The two common elements in all of these are (1) the alienation of people from nature and people from people, and (2) the compulsive resort to more powerful technologies and technological artifacts (electronic warfare, rock cocaine) to resolve the problems stemming from that alienation. Much like Humpty-Dumpty, the organic unity of individuals, social groups, and nature is hard to put back together again once it has been broken apart.

These observations are by no means original. Indeed, the connection between the contemporary ecological crisis and a reductionist world view has been a continuing theme in many areas of scholarly inquiry during the past few decades, including history of science, women's studies, environmental philosophy, and others. Holistic ideas have also had a practical impact in popular movements for self-health care, solid waste recycling, wilderness preservation, more balanced relations between the sexes, and a variety of other initiatives to transcend hierarchical and expansionist models of social relations.

132

Despite these promising developments, the political economy of advanced capitalism scarcely seems any less addicted to mindless growth today than it was at the time of the first Earth Day in 1970, so the problem of sustainability continues to loom large over humanity's horizon. Ironically, it is in the realm of political economy that humanity has gone farthest toward conceiving and constructing an integrated world system. Anyone concerned with the quality of human life on Spaceship Earth has much to learn from the captains of industry and state who created a global political economy, but ecologically their position at the forefront of global action puts the fox in charge of the hen house.

My premise in this essay is that the growth driven political economy of advanced capitalism is a fundamental obstacle to effective action on the various social concerns that arise from a holistic perspective. Issues like feminism, peace, and ecology should therefore be viewed as connected with the forces of production and the relations of authority; but in both intellectual and political terms, this is easier said than done.

As a consequence, most practical attempts to transcend the fragmentation of modern industrial existence leave the core relations of power and production untouched. Reforms like urban waste recycling or affirmative action typically ameliorate dysfunctions in the political economy without addressing the structures of accumulation that reinforce alienation from nature and among people. To put it simply, a society's system for dealing with its solid waste is important and reflects a broader strategy of ecological adaptation. But the strategy itself is shaped by the modes of relations of authority and forces of production available to a society. As two decades of environmental activism show, efficiency improvements in fuel economy or solid waste disposal can be negated in relatively short order by a political economy that promotes expanded consumption above all else. The political economy is a framework within which a society addresses its challenges and reproduces itself. It therefore needs to be a central focus of any effort to repair the breach between humanity and nature.[1]

A GLOBAL POLITICAL ECONOMY

In contrast to holistic ideas, which are very old, the articulation of social relations at a global level is of quite recent vintage. Social relations represent the interconnectedness of a global system in a new and profound way.

In the realm of political economy, finance as well as production and distribution have all transcended national boundaries, with finance leading the way. As Citibank chief executive officer Walter Wriston has expressed it, "Mankind now has a completely integrated international financial and informational marketplace capable of moving money and ideas to any place on this earth in minutes."[2] As a consequence, the exchange of foreign currencies now amounts to two hundred billion dollars per day, and the value of transborder capital flows is fifty times the value of trade in goods and services.[3]

The mobility of ideas and money is facilitated by the fact that these are commodities that can be exchanged in the form of electronic signals. In contrast, the physical components of industrial production are inherently more sedentary, a fact that constrains the integration of goods-producing industries at a global level. Manufacturing and other sectors that produce physical goods do not present the clear picture of a global system that can be observed in finance and certain service industries (e.g., tourism), but the restructuring of basic industry in recent decades nonetheless reflects some of the same underlying forces. For example, world trade (most of which is still comprised of physical goods) has grown nearly twice as fast as world output since 1965, which indicates the growing significance of global markets.[4]

Globalization has affected production as well as markets in the goods-producing sectors. It is increasingly common for the direct production of a single physical good to entail subprocesses at sites scattered across the face of the earth. The archetypal product of this system is the world car.[5] In aesthetics and technical characteristics it is designed for a global market. The means of producing it were similarly designed with an eye to the world as a single production site. The component processes of the global factory are terrestrially dispersed, while coordination and control are centralized in electronic space.

The globalization of industrial production is harder to measure in quantitative terms than the growth of currency exchange or trade. It is both conceptually and empirically difficult to distinguish finished products from components that cross national borders during a single manufacturing process. Furthermore, the various governmental agencies that compile economic statistics have made relatively little effort to do so. One exception is data on "coproduced" goods exported from the United States for processing in other countries (mostly Mexico and the Caribbean, but also South and Southeast Asia) prior to final assembly or shipping to the U.S. In the period from 1967 to 1983, for which data were compiled by the

United States Department of Commerce, the value of coproduced goods increased from $1.84 billion to $21.84 billion.[6] The manufacture of ideas has likewise gone global, with the proportion of scientific and engineering articles authored by persons of different nationality having tripled between 1973 and 1984.[7] New international consortia for industrial research and development are another important indicator of both the globalization of production and the key role played by scientific and technological innovation in this process.[8]

While the globalization of basic industry is hard to measure, and it is surely more constrained by geographic distances and political boundaries than finance and other services, its sources and consequences are fairly evident. By organizing production on a global scale, managers can take advantage of variations in the costs of factors such as labor, materials, energy, taxes, and pollution control, while gaining proximity to market niches. In addition to this extension of control over production and markets, the mobility of global enterprises provides a key advantage in bargaining with state authorities and local subcontractors over the terms of their interactions.[9]

To date, this globalization of the production process is observable in a relatively few industries, such as automobiles, semiconductors, textiles, and a handful of other manufactures and services. The defining technical characteristics of these industries are the conduciveness of the production process to spatial segmentation, and a high value-to-weight ratio that makes the transportation of components (which can be people in the case of service industries) economically viable.

The advances in communications and transportation, which were a primary requirement for global production, emerged from the post-World War II economic boom. Given the inevitable lag between technical advances and their incorporation into industrial practice, the earliest signs of globalization did not appear until the mid-1960s, with rapid industrial restructuring following through the 1970s and 1980s. The speed and the social impact of this restructuring have led many observers of these changes to predict a trend that will continue inexorably into the future. While critics view these developments darkly, technological cheerleaders applaud the globalization of progress.[10]

Both sides are wrong. Technical constraints, ranging from friction to market value, channel the globalization process down certain paths and preclude its entry into others, while the social dislocation associated with globalization generates political opposition. The future of the global political economy is, therefore, as much a matter of human values and social choices as of unconstrained technological growth.

Yet, even if globalization proceeded no further than it has at present, the conditions of humanity's material existence have been reconstituted by the emergence of productive institutions that transcend any single political jurisdiction on earth. Just as a small group of shareholders (five percent or so) can wield the balance of power in a large corporation, the global integration of key sectors of industry has fundamentally reshaped the conditions of production—and especially the political environment—for all industries.

The quantitative indicators presented above signal that something new is afoot in the world of production. However, the interconnectedness of local production with a larger whole that is the essence of globalism makes the distinction between the new and the old systems more conceptual than empirical. As Sharpston noted in an early study of international subcontracting, the key shift in the rise of global production took place in the minds of the managers and executives who conceive and execute the production process:

> There has been an increasing willingness among firms now called "multinationals" to think in terms of worldwide production and marketing possibilities, rather than treating overseas operations as separate from, and subsidiary to, operations in the country where such firms have their headquarters.[11]

Once managers started thinking of production in global terms, the reorganization of industrial practice was not long in following. In dealing with the political economy created in this image, we are likewise challenged to think of the whole world, rather than specific geographic locations within it, as a single production system.

Global production has not escaped the attention of academics and journalists. In fact, a virtual torrent of writings on the topic has flooded the media. However, this literature is permeated by the conceptual baggage of internationalism, which is quintessentially reductionist. The creators of global production may have provided new means of reshaping the human and natural environment, but political economy clings to old means of understanding the world.

GLOBAL OR INTERNATIONAL?

Our everyday lives are filled with signs of the new globalism. Those of us living in the Americas can rise in the morning to hear results of the day's events on the Tokyo Stock Exchange, or the value of the dollar in German marks, on virtually any radio or television station. What were previously reported as "foreign temperatures" in the daily newspaper's weather tables are now called "global" or "world" temperatures. New periodicals and broadcast programs focusing on "global" issues proliferate. Typical is *World Monitor,* a monthly publication of the *Christian Science Monitor.* An invitation for a charter subscription promises readers that they will learn of their commonalities with their "global neighbors," noting that many issues "transcend national boundaries, such as the environment . . . the family . . . education . . . and the flow of money."

The leaders of American foreign relations and industry created the infrastructure of the new globalism in the political and economic realms, but the rhetorical genre was invented by activists and dissenting scholars of the late 1960s who questioned the social and ecological implications of uninhibited industrial expansion. In time, established interests appropriated the rhetoric of globalism as well. By the late 1980s, even Ronald Reagan was speaking of the need for humankind to "live together peacefully on the planet Earth."[12]

The diffusion of globalist rhetoric was inevitably accompanied by confusion over exactly what it signified. The most common practice has been to use the terms *global* and international interchangeably.[13] Thus, while the rhetoric of globalism bespeaks the arrival of something new and different in human affairs, the language and underlying conceptualization often fall back upon the older concepts of internationalism that were presumably left behind with the advent of a global awareness. This blurring of new and old ideas can be conservative, because it tacitly confines thinking about power and production within the conceptual and historical framework of the nation-state system. State power, and the larger political economy and world view in which it is currently embedded, thereby becomes a limiting condition on inquiry about the human prospect.

This ambivalence about state power and global production is epitomized by the transnational enterprise (TNE). The TNE has pioneered the global integration of technology and commerce, but it is crucially dependent on the state system for the maintenance of the property relations, social order, and public goods (e.g., transportation infrastructure, education) essential to modern production. The TNE, therefore, promotes continuous change in technology and industrial organization at the same time that it resists political change that would undermine its privileged position in the state.[14] This leads to the paradox of a system that seeks "reform without reform."[15]

The relentless efforts of corporate image makers to associate the modern enterprise with all things new and progressive thus gloss over a steadfast opposition to popular initiatives that threaten class privilege, and they help to prevent the process of globalization from becoming politicized. One often hears of global markets and global competition in the business press and in the management schools where these are matters of everyday concern. An important function of this rhetoric is to frame global phenomena in the technical and commercial terms that underlay its original creation, while keeping politics off the agenda. Ambiguity about the nature and implications of global processes thereby serves established interests.

There are exceptions to this ambiguity among scholarly observers of global political economy. Political scientist Marvin Soroos, for example, defines global policies as a subset of international policies. They emerge from the whole international community (rather than subdivisions of this community, such as particular regions, rich countries) and deal with problems that are of "widespread concern."[16]

Soroos helps us see that the unique thing about globalism is universality and wholeness. Global processes deal with the whole world, while international processes can be as limited as the flow of traffic and other daily interactions between border cities. Both types of processes are important in today's world and will continue to be in the foreseeable future. It is the rapid growth in recent decades of global processes in culture that has caught us, conceptually and methodologically, by surprise.

When something pertains to the whole world, by definition it cannot be localized or isolated. In epistemology, holism refers to the embedding or enfolding of the whole into each part or region of a system. This seems to be what people intend by using the term *global* to refer to things like transnational financial networks, the greenhouse effect, and other problems that have universal implications.

These are new global issues, but things pertaining to the whole world have existed for as long as there has been one. Indeed, the very creation of the earth (whether one takes this to be an astrophysical or a divine event) was a global process. It was only later that human beings appeared on the scene, and later still that their activities began to have a global impact.

The contrast of physical with international processes is profound, for the latter have an identifiable starting point in human history and are therefore localizable in time and space. After all, before there could be inter-national processes, people had to create nations and nation-states. And just as people created international relations, they have the ability to terminate them, whether by means of a nuclear conflagration, a new form of political association, or some other alternative. Long after international relations are a distant memory—if indeed there is anyone to remember them—global physical processes will continue.[17]

This merely states the obvious but often forgotten: humanity is of the earth, not vice-versa. While humanity as a whole is embedded in the essentially unending story of our planet, human creations have finite lives, just like individual human beings. As a human creation, international processes are readily demarcated in time and space, but the difficulty of knowing the end point, since it is not yet upon us, obscures this fact.

Just as we reify international relations, we usually forget that global processes, relative to human society, are of infinite duration and usually exist quite independently of human agency. The indiscriminate use of globalist rhetoric trivializes the concept, but the new lexicon itself signals a profound development: the conversion of the world from which Homo sapiens emerged into an artifact.

Here, too, the new and the old coexist. Human societies have altered global patterns since they developed the capacity to move cultures over long distances. What is new is the mass production of global processes and the awareness of them that makes it possible to plan for global control.

To summarize, the sudden appearance of the human hand in global processes obscures several critical points: (1) international processes are human creations; (2) some global processes are human creations; (3) most global processes exist independently of human agency.

NATIONS AND NATURE

There are at least four reasons why muddled thinking on the nature of global processes is the norm. First, we live in a nation-state era and consequently take our status as citizens of particular countries to be the natural order of things rather than a historically specific experience. The concept of international relations flows readily from this frame of mind as the category for all encounters and interactions with entities outside one's national home. Yet the national boundaries demarcating the political loyalties of the earth's inhabitants have little relevance to many new developments in the cultural, economic, and technological spheres of the modern world, leaving us caught between a "second nature" born of the nation-state era and a world in which global interconnectedness is increasingly obvious.

Second, the human reconstruction of nature through science and technology has made it difficult to distinguish artifice from nature and has blurred the distinction between global and international processes. For most of human history, global events such as the coming and going of the Ice Age have been viewed as natural (or supernatural) ones, against which the human artifice paled by comparison. Science has undercut this tidy distinction by unveiling the worldwide impact of many human activities, while technology has provided the means of affecting nature at a global level. This makes it difficult to understand whether developments like the spread of industrialization to all corners of the earth are global or international in nature. The epistemological difference is a simple matter of wholes versus parts. As a practical matter, most people in most places on earth are involved in or affected by industrialization, which makes it a global phenomenon. But the industrial system would rapidly collapse without the public goods and order provided by the nation-state, which is the essential component of internationalism. When global meant nonhuman and international meant human, the distinction between the two was as clear as the difference between people and nature. But human societies have now reshaped the natural environment so profoundly that the old model no longer holds.

In the example just mentioned, my epistemological distinctions can be maintained by noting that industrialization is both global and international in terms of social organization and significance. Other issues do not permit this approach. This is especially true for visions of the human prospect, because they invariably hinge on some concept of the nature of human beings. For example, to most undergraduates at the technological university where I teach, the notion of exploring, inhabiting, and exploiting other planets is as natural as the child's curiosity about his or her body, or the teen's testing the limits of institutional authority. Many critics of big technology, on the other hand, view space colonization as an extension of the human artifice that literally separates people from their natural home on earth in an alienated pursuit of military domination and ceaseless growth in material output. The differences of perspective here are fundamental, and the genre of issues is of transcendent significance. The increasing association of human agency with global processes beclouds the epistemological ground on which artifice is distinguished from nature, thus permitting profoundly conservative and imperialist formulations of the human prospect to be put forth in the name of nature.

Once global processes have been reconstructed like an artifact, a small amount of epistemological slippage erases the distinction between humanity's bequest from nature and humanity's bequest from human creativity. Everything looks like it was made by people. The difference between international and global modes of social organization thus vanishes, reinforcing the presumption that international politics will eternally shape the global order.

The third dimension of confusion about global processes stems directly from the Western way of knowing. The particularist or reductionist science of Western society derives its power from the relentless isolation of parts from their context and the accumulation of ever more complete knowledge about their inner workings. The isolated part thereby becomes subject to the control of the scientist, and as knowledge of parts and their interconnections accumulates, social control over the total environment correspondingly expands.

The state too was conjured through this reductionist logic, as it literally appropriated power from individuals and prestate collectivities (most evidence indicates that simple bands fission when their membership exceeds twenty-five persons),[18] concentrating it in a formal institution that it created apart from the whole of society in order to rule over society. Like science, the state serves as a means of centralizing control over the environment, including the human environment. The early ascendancy of the state in the ancient Near East, where state construction and management of water works radically expanded human control over the environment in an arid climate, is no accident.

The modern nation-state emerged in the midseventeenth century after a long period of military conflict among the fiefdoms and monarchies of medieval Europe, putting to rest the final remnants of the long-decaying Holy Roman Empire. The most obvious result of this new institutional framework was a more stable political order that prepared the ground for the modern world. The more enduring consequences, however, are embedded in the very concept of nationality, which represents a particularist departure from the universalism of Rome.

A nation is a people who feel an affinity beyond the immediate ties of kinship or band. The basis of this belongingness is typically a unique collective creation, such as a language or ethnicity, and is generally associated with a homeland. The fact that some people belong to a nation means that others do not. Indeed, the word nation stems from the Latin *nasci,* "to be born," underscoring the fact that national entities have a uniqueness similar to that of individual human beings.[19]

The invention that combined nationality with statehood therefore joined institutional with intellectual power in fundamentally new ways. At the outset, more control over the environment was an obvious improvement over medieval warfare and stagnation. As the scale and ubiquity of the human presence in global processes has grown, however, the militarily and economically expansive characteristics of the nation-state have undercut its capacity to serve human needs. The global articulation of social forces requires ways of knowing and ways of being that are inclusive rather than exclusive, that emphasize openness and accept contingency, but the blurring of global and international processes traps us in the reductionist epistemology of the nation-state system out of simple habit.

In sum, reductionist epistemology shares some important features with the nation-state. Both were created to exercise power and control over people and nature. Both are intrinsically exclusive rather than inclusive. Both consign subjective concerns about justice and fairness to distinct, subordinate social spaces. And competitive gain—of status, power, and resources—is the modus operandi of both science and the state.

The final source of muddled thinking about globalism derives from the association of progress with technological change in Western thought. Most social inquiry on global matters celebrates the "progressive" character of the technological systems that have made it possible for human societies to affect things on a world scale. Global is good, whether it is the "world class" to which corporate managers and elite academics aspire, or the "planetary consciousness" of liberal activists. This association is programmed so deeply into the Western mind that one needs to virtually leave the culture in order to see that faith in progress and technological change is a superstition.

This does not mean that global processes are never cast in a negative light: terms like *global famine* and *world war* betray the darker meanings associated with big events. Much like science and technology, which also fall periodically from their pedestal in Western culture, the fascination with global processes and events can turn to fear at a moment's notice. But in both cases, thoroughly human activities are treated as autonomous entities. "Good" and "bad" qualities become inherent characteristics of these processes, rather than reflections of the human values and social choices that shape them. Given these epistemological underpinnings, the proclivities of particular cultures (for example, the generally positive view of technology and globalism in the West, or their association with "the great Satan" by Shiite Moslems) is a secondary matter. In either case, the central forces shaping the late twentieth-century world are represented as determining human action rather than resulting from it.

Those of us in the West can see all too easily that traditional cultures create problems for themselves by identifying modernity with the forces of evil, for the presumption that these forces are out of control undermines efforts to contain them. Clarity on this matter is ironic for two reasons: it deflects attention from the substantial harm inflicted on traditional peoples by the march of industrialization, and it conceals the greater deception wrought by the progressive ideology of the West. Western ideology blames the victims of progress for their own oppression, then covers its tracks by interpreting the globalization process through prepackaged media images and trendy scholarship, which provide the illusion of adaptation while leaving more fundamental patterns of thought and action largely undisturbed.

Taken together, the collective misperception of global developments reduces them to a spectacle that most people view from the sidelines, at once acknowledging change in the world without facing the discomfiting process of adapting to it. When global events are perceived to threaten human welfare (e.g., global temperature changes caused by industrial pollution), remedial efforts rarely acknowledge the distinguishing features of whole systems. Control is an aspect of global processes, embedded in their regular patterns of behavior, but the nonlocalizable character of global phenomena means that there are no leverage points from which to direct them. This runs against the grain of Western, reductionist thinking, which is perpetually in search of causal (controlling) variables in both inquiry and practical affairs.

HOLISTIC SOLUTIONS

There are everyday problems that people confront effectively in distinctly holistic modes. A good example is alcoholism. The alcoholic typically views his or her plight as a struggle of will against "the bottle." Yet the war against this external evil inevitably fails, for the longer one succeeds in abstaining, the stronger is the temptation to take "just one drink" and demonstrate conquest of the enemy. As all alcoholics know, the illusion of total control thus turns to total dependency and submission to the externalized bottle. The only way out of this cycle is to abandon the reductionist epistemology that isolates the bottle and the self as separate elements doing battle and to instead own up to being a drunk. Control over one's life then becomes a choice between being a ruined alcoholic or a recovering alcoholic. Not being an alcoholic, however, is not one of the choices. The addictive aspect of the alcoholic's being cannot be separated from the remainder of the self and controlled, because the self is the totality of interactions between each individual and her or his environment.[20] People engage in all manner of contorted behavior to get outside of this totality, but the very idea makes about as much sense as the childhood fantasy of a place outside the universe.

The pragmatic holism that enables alcoholics to reformulate their senses of self in the world has implications for global processes in general. This is no place for a blueprint, but I can at least sketch out some directions for thinking about problems embedded in global production. The key, as for the drunk, is acknowledging that control comes *in* situations, not over them.

Consider the problem of drug abuse. The production of illicit drugs is a global matter on three counts. First, ownership and management is global in scope and therefore can move from one nation or region to another when state authorities create adverse conditions. Second, modern agribusiness has dislocated traditional agriculture in all parts of the world, creating a social and economic void that has been filled by export agriculture, the most lucrative crops being marijuana, coca leaves, and opium poppies. Finally, the alienation of modern urban existence has encouraged various forms of escapist behavior, including habitual drug use, creating a market for illicit mind-altering substances in capitalist, socialist, and Third World countries alike.

The ubiquity of both the demand for illegal drugs as well as the supply of factors needed to produce them (capital, labor, and land) deprives territorial authorities of levers for controlling drug abuse. The primary objective of drug abusers is loss of personal control and responsibility in the world, and the global drug trade is efficiently organized to support this objective. The contrast with traditional narcotics use is the contrast between traditional and global societies. Indians of the highland Andes, whose use of the coca leaf goes back for centuries, are shocked by the notion that anyone would magnify its narcotic effect by converting it to rock cocaine. But under capitalism, addiction is a source of profit, and globalization places the industry beyond the control of state authorities. What has emerged is a destructive and ubiquitous combination of mobile capital, willing producers, alienated users, and powerful new drugs that make it possible to function in mainstream society while under the influence. The very wealth of modern society offers an array of opportunities for financing an addiction, ranging from high income jobs to careers in theft and prostitution. The illegal options assure access to drugs regardless of social class and education.

Just like the drunk, global society can not get outside of itself. The war on drugs, more than most wars, will be remembered chiefly for the amount of blood that it spills.[21] I make no claim that getting outside this solipsistic loop will be easy, whether the issue at hand is drug abuse, global climate change, immiseration of the Third World, or the threat of nuclear warfare. I feel reasonably confident, however, that our fundamental choices turn on what kind of global society we will be, rather than whether or not to be one. Until we can grasp the extent to which production and other social processes have been globalized, and the differences between global and international forms of social organization, we will be unable to address the ethical and political challenges of the world we have created.

NOTES

I would like to thank Joe Brown and Shirley Gorenstein for their close reading and critical comments on an earlier draft of this paper. Discussions with Michael Black were critical to the basic conceptualization.

1. In a previous paragraph I listed more equal relations between the sexes as one practical gain associated with the revival of holism, but the implications of this development are surely not as limited as the other examples I cite. Just as a political economy shapes the reproduction of social institutions, relations between the sexes shape the reproduction of individuals, in the biological and social senses. I think of political economy and sexuality as domains of social activity in which regenerative (and in that sense primary) processes are concentrated.

2. Michael Moffitt, *The World's Money: International Banking from Bretton Woods to the Brink of Insolvency* (New York: Simon and Schuster, 1983), 43.

3. Warren Bennis, "Managing the Dream: Leadership in the 21st Century," *USC Business* 1 (1990): 35.

4. Calculated from tables in World Bank, *World Development Report* (New York: Oxford University Press, 1987).

5. Kurt Hoffman and Raphael Kaplinsky, *Driving Force* (Boulder, Colo.: Westview Press, 1988).

6. Joseph Grunwald and Kenneth Flamm, *The Global Factory: Foreign Assembly in International Trade* (Washington: Brookings, 1985), 13. Grunwald and Flamm also report the growth of similar arrangements between Japanese companies and subcontractors throughout Southeast Asia.

7. National Science Board, *Science and Engineering Indicators, 1987* (Washington: National Science Foundation, 1987), 285.

8. Herbert I. Fusfeld, *The Technical Enterprise: Present and Future Patterns* (Cambridge: Ballinger Publishing Company, 1986), 127–40.

9. For more detailed analyses of these interactions, see Michael Black and Richard Worthington, "Democracy and Disaster: The Crisis of Metaplanning in America," *Industrial Crisis Quarterly* 2 (1988): 33–51.

10. Technological optimists who address these issues include Robert Gilpin, *The Political Economy of International Relations* (Princeton: Princeton University Press, 1987); Judd Polk, "The New International Production," in *Contemporary Perspectives in International Business,* ed. Harold W. Berkman and Ivan R. Vernon (Chicago: Rand McNally, 1979); and Simon Ramo, "The Foreign Dimension of National Technological Policy," in *Technological Frontiers and Foreign Relations,* ed. Anne G. Keatley (Washington: National Academy Press, 1985). A good collection of critical essays is edited by Edward W. Gondolf, Irwin M. Marcus, and James P. Daugherty, *The Global Economy: Divergent Perspectives on Economic Change* (Boulder, Colo.: Westview Press, 1986). See also Harley Shaiken, *Work Transformed: Automation and Labor in the Computer Age* (New York: Holt, Rinehart and Winston, 1984).

11. M. Sharpston, "International Sub-contracting," *Oxford Economic Papers* (March 1975): 99.

12. Ronald Reagan, "Managing the Global Economy" (U.S. Department of State, Bureau of Public Affairs, Current Policy Series no. 1006, text of a speech delivered before a joint meeting of the World Bank and the International Monetary Fund, Washington, D.C., 29 September 1987).

13. For example, the authors of a recently published marketing textbook use *international* three times, *global* twice, and *worldwide* once in a single paragraph, all with the same meaning. See Jagdish Sheth and Abdolreze Eshgi, *Global Marketing Perspectives* (Cincinnati: South-Western Publishing Co., 1989), vii.

14. My terminology here is a slight variation on Lindblom's theory of the "privileged position of business" in polyarchies (industrial democracies). See Charles E. Lindblom, *Politics and Markets: The World's Politico-Economic Systems* (New York: Basic Books, 1977). Although globalization has changed state-society relations, business retains a privileged (and I would argue enhanced) position. See especially Susan Strange, "Toward a Theory of Transnational Empire," in *Global Changes and Theoretical Challenges: Approaches to World Politics for the 1990s,* ed. Ernest-Otto Czempiel and James N. Rosenau (Lexington, Mass.: Lexington Books, 1989).

15. Alan Wolfe, *America's Impasse: The Rise and Fall of the Politics of Growth* (New York: Pantheon, 1981), 80–102.

16. Marvin Soroos, *Beyond Sovereignty* (Columbia: University of South Carolina Press, 1985), 20.

17. There is always the possibility that an astrophysical event will simultaneously destroy the planet and its inhabitants, but this possibility does not alter the fact that human processes are embedded in a global system.

18. Joseph Birdsell, "Some Predictions for the Pleistocene Based on Equilibrium Systems among Recent Hunter-Gatherers," in *Man the Hunter,* ed. Richard B. Lee and Irven De Vore (Chicago: Aldine Publishing Company, 1968).

19. Robert Lopez, *The Birth of Europe* (New York: M. Evans, 1963), 94–108.

20. Gregory Bateson, "The Cybernetics of 'Self': A Theory of Alcoholism," *Psychiatry* 34 (1971): 1–18.

21. The burden on the criminal justice system and the number of people incarcerated as a result of this war are also astounding. In California, drug crimes and drug related offenses (e.g., burglary to support a drug habit) account for 80 to 85 percent of all criminal cases in metropolitan areas. Drug law violations more than tripled statewide in the decade from 1978 to 1987, while the prison population quadrupled. See Philip Hager, "Drug Cases Now Dominate Caseloads of Urban Courts," *Los Angeles Times,* 23 May 1990.

No Innovation Without Representation: Technological Action in a Democratic Society

STEVEN L. GOLDMAN

When the American Colonists made "No Taxation without Representation" a rallying cry in the 1770s, they were claiming an established right as Englishmen to participate in parliamentary decisions directly affecting their well-being. The operative term in this slogan is *representation*. The colonists were not demanding veto power over their taxation, only the right to participate on the same terms as other Englishmen in the tangled political process— the debates, the negotiation of competing interests, the exchange of favors—that determines the fate of legislation. For it must be appreciated that taxes do more than take cash out of the pockets of citizens. By differentially affecting the costs of goods and services, tax laws influence decisions to manufacture products or to import them, the status of raw materials, the need for education and technical expertise, the choice of markets for all of these, the distribution of economic and political power within a society, and the relations of a society with other societies. Colonial taxation policies were correctly perceived by the colonists as the imposition upon them of a very particular future envisioned for them by Parliament in London.

Today, science and technology policies have a social impact comparable to that of taxation policy in the colonial period. In 1776, political freedom entailed the right to a voice in taxation decisions because these decisions were primary forces shaping the fabric of personal and social life. As the power to make taxation decisions belonged to Parliament, the colonists demanded representation in Parliament. Today, political freedom entails the right to representation—again, not veto power, but meaningful representation—in corporate and governmental policy decisions; to pursue

148

particular lines only of scientific and engineering research and development; or to encourage by direct and indirect subsidy the implementation of certain new technologies rather than others and in particular ways. Freedom requires this because scientific and technological action decisions are acknowledged on all sides as profoundly affecting the fabric of our personal and social lives.

For the last thirty years at least, in all of the technologically developed nations and in all of the rapidly developing nations, the same mantra has been recited over and over: "In technology lies the future." But if the technologies that are implemented shape the future that we will have to live, then it seems compelling to ask *which* technologies, and therefore *which* future, are we to implement? And *whose* technologies are these to be, which is as much as to ask, to whom shall the future belong? In fact, however, these questions are not being pressed by the American public today, nor is there a widespread demand for public representation in the bodies making these decisions, so pregnant with personal, social, political, and economic consequences.

One reason for this apathy is that these decisions are perceived as belonging to corporate and governmental agents of technological action. The public is excluded from the former on the ground that entrepreneurial decisions are properly proprietary in a capitalist society. They are excluded from the latter on the twin grounds that the public lacks technical expertise, and that their interest is already represented, by definition, in governmental agencies. But if science and technology policy decisions are always political decisions, then, in a democracy, the public has a right to participate in these, as in all political decisions regardless of where they are made. Furthermore, technical expertise is not now a condition of deciding corporate or political science and technology policies, and current representation of the public's interest—for example, in federal regulatory agencies—is inadequate because it is a reaction to decisions that have already been made, or to implementations that are already in place or underway.

A second reason for public apathy on science and technology policy decisions is a pervasive misconception of the nature of technological action. Technological innovation, to paraphrase David Noble, is neither automatic, nor Darwinian, nor objective.[1] The course of innovation is shaped primarily by the subjective values and the contingent goals of technological actors. It is to these actors and their motivations that we must turn if we want to understand new technologies, not just to the availability of new bodies of knowledge. The belief that innovation is an essentially objective process driven by the technical knowledge of scientists and engineers disguises the actual subordination of technical knowledge to the institutional dynamics of technological action. Innovation has been an authoritarian process at least since the steam power based Industrial Revolution began, precisely because innovation always implicates vested interests. Innovation has been, and promises to continue to be, a source of personal and social wealth, power, privilege, and prestige. It is hardly surprising then, that the public, intimately affected though it knows itself to be, stands outside the innovation process, which is presented today as it was at the Chicago International Exposition of 1933. The motto of that exposition was: "Science Finds, Industry Applies, Man Conforms!" Scientists, engineers, and industrialists act; the public reacts.[2]

Technological action is a social process in which scientists and engineers participate, but which they neither initiate nor control.[3] The forces driving that process are initiated, and constantly modulated, by context specific, hierarchically structured, "managerial" decisions. These decisions selectively apply technical knowledge to the accomplishment of the agendas of the institutions the managerial decision makers serve. I use the term *managerial* to emphasize the extent to which technological action is determined by considerations of a highly contingent character. These considerations are interpretations by people with decision-making authority—in a company, corporation, venture capital firm, or government agency—of what the exploitation of technical knowledge in one way rather than another can achieve for them, given the institutional, personal, and societal influences to which they are consciously and unconsciously responsive.

Illustrations of this abound. That, without exception, all of the currently functioning commercial nuclear power plants in the United States are light water reactors and of a *de facto* standard size (for those ordered since the late 1960s) of one thousand to twelve hundred megawatts, is not at all a reflection of the only reactors that nuclear engineers could have built, or would have built, had the choice been left to them. The total dominance of light water reactors in the United States is a consequence of the involvement of General Electric and Westinghouse in the naval nuclear submarine program, the upshot of which was that these companies acquired experience with, and government funded facilities for manufacturing, light water reactors. The size of reactors built since the late 1960s is a consequence of General Electric and Westinghouse attempts at recouping the losses they suffered on the five hundred to six hundred megawatt plants that they offered utilities at "loss leader" prices in the early 1960s in order to precipitate a commercial nuclear power industry.[4]

The very existence of a commercial nuclear power industry is itself a consequence, neither of the nuclear science and engineering knowledge bases available in the 1950s, nor of a perceived need on the part of U.S. utilities for alternatives to coal, oil, and natural gas, but of deliberate federal policy decisions. For economic, political, foreign policy, and national security reasons, the Truman and Eisenhower administrations, the Atomic Energy Commission, and the Congressional Joint Committee on Atomic Energy were determined to commercialize nuclear technologies. At the same time, managerial decisions within the Atomic Energy Commission shaped the developing nuclear science and engineering knowledge bases by the types of power reactor and nuclear weapons research and development projects they funded, by the production and testing processes they supported, the safety and reliability studies they commissioned, and the standards they set.[5]

The fusion power program in the United States has followed a similar course, as Joan Bromberg has documented.[6] The now thirty-five-year long commitment of many billions of dollars of federal funds to commercial fusion power research and development is itself a political decision of considerable consequence. Furthermore, the history of this commitment and the current state of the fusion power art reflect the repeated imposition of political, and military, considerations on ostensibly objective technical research programs. This imposition is embodied in managerial appointments to the federal agencies managing the program, currently the Department of Energy, and in the priorities for the program set by that agency. It is reflected in the changing allocation of funds: among competing reactor designs; between reactor science and reactor engineering; between research, design studies, or construction of experimental or prototype machines of different types; among competing federal, academic, and private laboratories. The result is the selective accumulation of technical knowledge whose character and distribution are to a large degree expressive of fluctuating subjective value judgments by successive presidents and congresses.

The Apollo space program rested on a personal political decision by President Kennedy.[7] The political astuteness of this decision was amply demonstrated by the overwhelming national and international support it received at the time. But the upshot was that our concern with space has ever since been driven by political judgments. We have never really had a civilian space policy, and so our space initiatives have lacked coherence and continuity. The Apollo program went to the moon, and then nowhere. Spacelab literally fell out of the sky. The design of the Space Shuttle was dramatically affected by political compromises made by NASA with Congress and the Department of Defense in order to win funding for it.[8] Those compromises weigh on all our current space activities and limit our mission options for the next twenty years at least.[9] The design of the space station *Freedom* is today undergoing similar modification as NASA, Congress, and the Bush administration negotiate what can be built, for how much money, seemingly independent of any rationale other than national prestige.[10]

The pervasiveness of the computer in American society today is substantially the result of the introduction of the IBM PC in the fall of 1981. But the success of the IBM PC came in spite of the fact that it was less technically advanced than a number of micro-computers then available. It succeeded, first, because corporate computer buyers saw through the machine itself to what the IBM corporate logo symbolized. Corporate managers saw in the IBM product a commitment to a long term standard that would not embarrass corporate buyers by a rate of technical advance that would make their computer purchases obsolete every six months. And it succeeded because IBM management, in a dramatic company policy reversal, deliberately opened the PC to third party vendors of hardware and software products. Eight years after its introduction, the IBM PC-type of computer is still the dominant personal computer, and the tens of millions of them in daily use suggest that it will survive a lot longer, indifferent to the continuing advance of computer technology. For much the same reasons, the Apple II continues to generate a substantial fraction of Apple revenues, even though it has a high price and is technically inferior to Apple's Macintosh computer.

Indeed, it is not much of an exaggeration to say that the existence of the Apple Corporation itself, and of the IBM PC, were the result of managerial decisions to reject new technical options. Hewlett-Packard management decided not to accept the offer of Steve Wozniak, then an employee, to manufacture his prototype micro-computer as a Hewlett-Packard product.[11] At Digital Equipment Corporation, chief executive officer Ken Olson rejected a proposal in 1978 to configure a DEC PDP-8/F minicomputer as a personal computer because he could not see why anybody would want to own their own computer.[12]

In *The Soul of a New Machine*, Tracy Kidder documented the

creation by Data General Corporation engineers of that company's first thirty-two-bit computer. They had to work under management imposed constraints that deliberately restricted the performance, and the technical "elegance," of their new machine to ensure consistency with corporate commitments and nontechnical policy objectives.[13] Peter Temin's account of the breakup of the Bell system clearly reveals the impact of managerial policies on Western Electric engineering, as well as on the research program of the Bell laboratories.[14] Margaret Graham's history of the RCA videodisc project, from its inception in 1965 to its termination in 1981 (with losses well in excess of 500 million dollars), similarly reveals the domination by management decisions of the content, the effectiveness, and the application of RCA's corporate technical resources.[15] Defense Department weapons system research and development provides too many illustrations of this phenomenon even to list, but among the most recent and egregious examples are the B-1 (and perhaps the B-2) bomber, the Los Angeles-class nuclear attack submarine, the Trident II SLBM, and the Strategic Defense Initiative. In each case, the final "high tech" product reflects politically motivated decisions and compromises far more than an application of the best available technical knowledge, commonly at the expense of the very mission that the weapon system was initially intended to perform.[16]

The implications of a characterization of technological innovation as a managerial, decision dominated rather than a technical, knowledge dominated process are profound. The dynamics of the innovation process are to be located, on this view, in the institutional contexts of technological action, not in engineering or scientific expertise. Overwhelmingly, today, these contexts are either large corporations or federal governmental agencies. The practice of science and engineering in the United States typically is embedded within the activities of large-scale enterprises employing at least hundreds, and often thousands, of scientists and engineers, among many thousands of additional employees.[17] The orderly functioning of these enterprises necessitates highly elaborated hierarchical managerial structures. The function of this hierarchy, no matter how high the level of technology being exploited, is not to serve the enthusiasms of the scientists and engineers, but the reverse: for scientists and engineers to assist in accomplishing management's agenda on management's terms.

Furthermore, managers do not simply do what they are explicitly told to do by their superiors. The management of a large company would be almost impossibly unwieldy if that were the case. To be effective, managers need to adopt as their own the objectives set by their superiors. Sometimes this requires setting aside—because of cost considerations, or scheduling conflicts, or performance commitments, or stylistic requirements—what technically trained managers might have thought of as prudent engineering, had they still been engineers. One thinks, in this regard, of the Ford Pinto fuel tank problem and the General Motors X-car rear brake problem; also of the Morton Thiokol engineers who tried to postpone the Space Shuttle *Challenger* launch on technical grounds. The Thiokol engineers were told by their superior to take off their engineering "caps" and to put on their management "caps."[18] This change of perspective was expected to cause them to subordinate narrow engineering concerns to the corporate concerns of Morton Thiokol and so approve the launch in keeping with the apparent wishes of a very valuable customer, NASA.

The logic of any technological action decision, then, is specific to its institutional context. That is, each decision makes sense only in relation to its institutional context at the time that it was made. The tendency to attribute to technical expertise primary responsibility for the products of technological action is understandable but misleading. It has all the appeal of substituting simple appearances for a complex reality. In the process, this reinforces the power and the authority of managerial decision makers by obscuring the value laden nature of the role they play. Appreciating that the *locus* of the selectivity that determines the course of technological action is contingent managerial decision making—rather than objective limits of technical knowledge, or objective performance standards, or market forces—decisively undermines popular notions of technology and technology management as objective enterprises.[19] It opens the possibility that innovation is a political process because managerial decisions always have a political dimension to them.

Technological action is always political action because it is always a form of action on a *polis* that draws its self-serving rationale from the institutions and values of just that *polis*. It is always action whose selective application of technical knowledge is calculated to advance the technological actor's interests within that *polis* by exploiting prevailing values and their institutionalization.

Technological action is thus the end product of a series of value judgments in which society is implicated from beginning to end. Society is the explicit object of that action, and social criteria determine the range of possible courses of technological action. At the same time, the social context of technological action, and the constellation of political, economic, legal, and cultural institutions and values extant at a given moment in a society's existence, shape the managerial agendas that determine the applications of technical knowledge. Technology is therefore inseparable, not merely from the concrete institutional contexts of its practice; it is inseparable as well from the broader social environment of which those contexts are themselves expressions. That is, institutional technological actors do not unilaterally impose their corporate wills on society. Corporate and governmental entities and their agendas are themselves social phenomena. In the broadest sense, technological innovation is always a process in which a society acts on itself. The mechanism of this action, however, is the selective manipulation of institutions and values by individuals, either on their own behalf, or on behalf of parochial interests they serve.

Paradoxically, one corollary of this realization is that managerial actors do not, in fact, control the ultimate social impact of technological innovations. The reason why this is so is that any body of technical knowledge is much richer in potential applications than any selective exploitation of it. Part, at least, of the unpredictability of the course of technological innovation surely derives from this fact.

A second corollary is that if a society wishes to affect the social impact of innovation, then the most effective means of doing so is by acting on its own institutions and values, on which courses of action are or are not going to be permitted in that society, and at what cost. It does not need to act on bodies of technical knowledge, nor does the public in such a society need to acquire technical expertise in order to play a role in technology oriented managerial decision making. This should be self-evident. Members of Congress, presidents and their cabinet secretaries, the military high command, and corporate management in the overwhelming majority of companies, have either no training or no current competence in science or engineering. Nevertheless, these people, advised by their own technical experts, routinely set science and technology policies for their institutions. Excluding the public from these questions is strictly a matter of maintaining power relationships. If the will to do so existed, the same technocratic infrastructure that services high-level political, military, and corporate decision makers could provide their putatively value-free technical expertise to the public as well.

The purported value neutrality of the technical appears to be an ideologically motivated stratagem. It serves to insulate from criticism the dynamical factors actually determining technological action. Technology, after all, is directly implicated in changing the world. It is, therefore, directly implicated in the struggles for power among the institutions and individuals in society who see threats to their interests in new technologies, or who see in new technologies opportunities for expanding those interests. One consequence of promoting technological action as driven by value-neutral technical knowledge is that the management of powerful vested interests is simplified by the public's not appreciating the role played by management in selectively creating and applying technical knowledge. Were they to appreciate this, the public would more readily ascribe to management, rather than to the inherent unpredictability of technology, direct responsibility for the social consequences of particular technological applications.

If managerial decision making, reflecting parochial institutional agendas, is the key to technological innovation and its social impact, how can the public hope to participate in the process in any meaningful way? This decision making is, after all, proprietary. It is insulated from interference by "outsiders" not only in capitalist but also in socialist and even in communist societies. It is protected in the former in the name of the sanctity of private property; in the latter, in the name of efficiency and specialized knowledge, or by a "people's" bureaucracy as impenetrable to its citizenry as Fort Knox is to the public whose money it holds.

In principle, the questions bearing on innovation that are the most obviously political are easy to ask. Should the United States have an innovation policy? If so, who should set it and on what grounds? Should we, for example, take action to generate and/or reallocate resources to pursue current competitiveness in international trade, or future competitiveness, or the greatest good for the greatest number of our citizens? Should we pursue self-reliance or international interdependence? What weight should be given to conservation and to low environmental impact? How are we to define the public interest and how are we to prioritize competing and conflicting claims by different segments of the public?

These are easy questions to ask, but no means currently exist for politically efficacious public consideration of them. Instead, innovation activity today is localized either in intensely parochial commercial settings, or in politically motivated government agency settings. Innovation is highly dependent on managerial decisions insulated from external, public based considerations that are not imposed by law. The automobile industry can argue the impossibility of meeting corporate average fuel economy requirements, the inefficacy and/or economic impossibility of emission controls, air bags, eye-level rear brake lights, or bumper standards until legislation is passed, after which the impossible is achieved and even exceeded. The petroleum industry, faced with imminent restrictions on automobile use because of air pollution levels, suddenly discovers that new gasoline formulations that dramatically reduce pollution not only are possible but *now* can be made available and are economically feasible.[20]

Given this situation, what can be done to create a niche for the public in the innovation process? First, and most important, the public needs to understand the fundamentally political character of innovation. That is, the public needs to understand that claiming a role for itself in the innovation process is neither left-wing nor anticapitalist, but a corollary of democracy. A people cannot be said to govern themselves if they have no say in the initiation of actions of profound consequence for their personal, economic, and political relationships and values.

Second, having claimed such a role, teaching innovation as a social process must be incorporated into the universal education system so that all of the public will be exposed to it as part of citizenship training. This will require distinguishing engineering from technology, with which engineering is often erroneously identified, in order to highlight the role of extratechnical value judgments in innovation. A by-product of bringing this into the primary or early secondary school curriculum will be the opportunity to introduce students at an early age to engineering as a profession distinct from science, a profession with the largest number of practitioners in the United States and one that plays a significant though misunderstood role in innovation.[21]

Third, the public needs to appreciate that the current regulatory, legislative, and judicial means of affecting the innovation process, while valuable, are not enough. Laws, lawsuits, and almost all regulatory agency actions are responses after the fact to innovation decisions made without public participation. Furthermore, these responses generally follow not merely decisions to act, but the expenditure of millions and even billions of dollars in facilities. This creates a climate that severely limits the scope of public reaction in all but the most egregious of cases. Once a utility has expended a billion dollars on a nuclear power plant, it is very hard to forbid its completion. Once a product is in production, it is very hard to ban it. Once a weapon system has been developed, it is very hard not to deploy it. But prospective regulation, which has been the hallmark of the nuclear power industry, and more recently of the commercialization of genetic engineering techniques, has not been successful either. As with retrospective regulation, initiatives for developing new technologies, for the commitment of capital, and the anticipation of profit are kept separate from the regulatory and legislative processes, creating either an adversarial environment for innovation or a manipulative one.

At the same time, it should be obvious that the marketplace, too, is not a substitute for direct public involvement in the managerial decision-making phase of innovation. The market would not be a satisfactory mechanism even if it were truly free, which to a discouraging degree it is not, because a large segment of the market is not competitive at all. The agriculture sector, for example, is wholly entangled with federal and state price supports, subsidies and price regulations. The defense sector of the economy, the space, research, and nonprofit sectors are also not subject to public response as a means of expressing public pleasure or displeasure, while many segments of the consumer market are dominated by a small number of manufacturing, distribution, and retailing companies.

Finally, what is needed for the public to play a role in the innovation process commensurate with our rhetoric of being a free people is the creation of mechanisms for introducing the public's interest before the fact, for participating directly in the making of managerial decisions rather than attempting to undo them, or to modify them, after they have been made. What form these mechanisms will take is not yet clear, although it is easy to think of mechanisms that are unlikely to work. At this time, the appropriate course of action seems to be to provoke public opinion, debate, and activism, in anticipation of constructive proposals emerging. It would be an accomplishment to precipitate broad debate on what the public interest is in any particular issue, how it might be articulated, and who might legitimately speak for the public. In the meantime, much more of an effort needs to be made to communicate the legitimacy of public activism in this regard and to intensify, through the existing after-the-fact mechanisms, corporate and governmental agency accountability to the public for their innovation decisions.

One course of action seems least likely to stimulate better means of representing the public interest in science and technology policy making: easing regulatory, legislative, and judicial pressures on technological actors. However imperfect these may be, and however inadequate or even distasteful, only through the pressure they bring to bear do we create incentives, among both public interest activists and corporate managers, to identify new mechanisms that will be more perfect and less distasteful.

It is important to recall that most of the mechanisms now in place, from the Environmental Protection Agency and the Occupational Safety and Health Administration to Ralph Nader's activism and Jeremy Rifkin's law suits, are products of public pressure that grew to politically significant proportions in the late 1960s and early 1970s. This opened the door to public involvement in innovation and research decisions, albeit after the fact. We now need to build on this foundation by using it as a foundation, applying the pressure these mechanisms allow to provoke new mechanisms that will allow public representation before the fact. We may not end up with wiser decisions, but it is the only course of action open to a free people.

NOTES

I am indebted to Professor Lew Fikes, director of Ohio Wesleyan University's National Colloquium series, who invited this essay as a contribution to that series and with whose permission it appears here. A lecture based on this essay was presented at Ohio Wesleyan on 15 November 1989.

1. David Noble, "Automation Madness, or the Unautomatic History of Automation," in *Science, Technology and Social Progress,* ed. Steven L. Goldman (Bethlehem, Pa.: Lehigh University Press, 1989), 65–92.

2. For an extended discussion of the ideology of the Chicago Fair's motto and iconography, see John M. Staudenmaier, S.J., "Perils of Progress Talk: Some Historical Consideration," in Ibid, 268–97.

3. Steven L. Goldman, "The Social Captivity of Engineering," in *Critical Perspectives on Engineering and Applied Science,* ed. Paul Durbin (Bethlehem, Pa.: Lehigh University Press, 1991).

4. Irvin C. Bupp and Jean-Claude Derian, *Light Water Reactors: How the Nuclear Dream Dissolved* (New York: Basic Books, 1978).

5. Ibid., 15–41, 53–55; also Stephen Del Sesto, *Science, Politics and Controversy: Civilian Nuclear Power in the United States, 1946–1974* (Boulder, Colo.: Westview Press, 1979).

6. Joan Lisa Bromberg, *Fusion: Science, Politics and the Invention of a New Energy Source* (Cambridge: MIT Press, 1983).

7. John Logsdon, *The Decision to Go to the Moon* (Chicago: University of Chicago Press, 1970); and Walter McDougall, . . . *the Heavens and the Earth: A Political History of the Space Age* (New York: Basic Books, 1985).

8. John Logsdon, "The Space Shuttle Program, A Policy Failure?" *Science* 232 (30 May 1986): 1099–1105; also Thomas H. Johnson, "A Natural History of the Space Shuttle," *Technology in Society* 10, no. 4 (1988): 417–24.

9. *Round Trip to Orbit: Human Space Flight Alternatives,* a report to the Congress of the United States by the Office of Technology Assessment (Washington, D.C.: USGPO, 1989).

10. Sylvia D. Fries, "*2001* to 1994: Political Environment and the Design of NASA's Space Station System," *Technology and Culture,* 29, no. 3 (July 1988): 568–93.

11. Paul Freiberger and Michael Swaine, *Fire in the Valley* (Berkeley: Osborne/McGraw Hill, 1984).

12. Ibid., 20.

13. Tracy Kidder, *The Soul of a New Machine* (New York: Avon Books, 1981).

14. Peter Temin, with Louis Galambos, *The Fall of the Bell System* (Cambridge: Cambridge University Press, 1987).

15. Margaret B. W. Graham, *RCA and the VideoDisc: The Business of Research* (Cambridge: Cambridge University Press, 1986).

16. See, for example, Nick Kotz, *Wild Blue Yonder: Money, Politics and the B-1 Bomber* (New York: Pantheon Books, 1988); and Patrick Tyler, *Running Critical: The Silent War, Rickover and General Dynamics* (New York: Harper and Row, 1986), especially 154–58.

17. *Engineering Employment Characteristics* (Washington, D.C.: National Academy of Engineering Press, 1985).

18. *Report of the Presidential Commission on the Space Shuttle Challenger Accident,* William P. Rogers, chairman (Washington, D.C.: USGPO, 1987), 93.

19. David Noble, "Automation Madness."

20. The profits of the Fortune 500 largest U.S. industrial corporations for 1988 are published in *Fortune* magazine for 24 April 1989. "Profits in constant dollars exceeded those of the previous inflation-adjusted record year, 1980. Any way you cut it, last year was the most prosperous for the 500 since *Fortune* published the first 500 directory in 1955" (p. 340). For the first quarter of 1989, after tax earnings for 664 "major corporations" increased 11 percent (*Wall Street Journal* Monday, 8 May 1989, A5). A comparison of U.S. corporate profitability with that of Fortune's International 500 directory (*Fortune,* 1 August 1988) is very revealing in relation to America's industrial competitiveness "crisis."

21. Steven L. Goldman, "Philosophy, Engineering and Western Culture," in *Broad and Narrow Interpretations of Philosophy of Technology,* vol. 7 of *Philosophy and Technology,* ed. Paul Durbin (Dordrecht: Kluwer, 1990), 125–52.

Part 3
New Issues

Introduction: New Dimensions for Action

ELENA LUGO

Contemporary scientific knowledge and technological know-how provide human beings with a new dimension for action on the world, with a more extensive, as well as a more intensive, power to control and to modify nature, including human nature, and to design artificial entities. Because modern conceptions of the world do not envision it as a divine creation, but rather as an autonomous, self-sufficient universe, citizens of modern societies withdraw from any bonds with a deity and deny divine intervention in their affairs, while attributing to themselves self-sufficient power over their existence. Autonomous reason, conscious of its own power, encourages us to interpret and to intervene in most, if not all, areas under our awareness according to strictly immanent laws. Transformation of self, and not merely becoming what we are by essence, inspires our endeavor. Self-liberation by way of scientific technical knowledge is thus now conceived to be possible, but in fact it lacks foundation in the orders of meaning, of being, and of values. So, we must now ask, is this new world, which is necessarily alien to anthropological, ontological, and axiological principles, a proper context for a truly human existence?

Modern men and women are not content with enough technological control of the world to satisfy their basic and vital needs. They seem to want to penetrate the world itself with a new pragmatic spirit and a sense of their domination of it, so as to experience and represent themselves as lords and masters. They not only seek to make matter useful to them according to their needs, but they also seek to bring out, creatively, a new intelligibility (instrumental and artificial) in matter. Technology thereby assumes a symbolic character and function within a technology defined social and political context, by modifying the self-conception of human existence, the temporal and spatial dimensions of human existence, and of human bonds to nature and to one another.

Thus, the intensity of technological activity, the magnitude of its cosmic influence, the capacity to generate new forms and artificial entities as well as its accumulated achievements, and the irreversibility of its consequences, are new factors that must be considered in any ethical appraisal of our social responsibility toward technology. As Hans Jonas has repeatedly claimed, the chasm between our technical capacity and our ability to predict the future consequences of its application requires ethical prudence and caution as primary duties.[1] We need moral fortitude to guide our technical power. We must also recognize that the responsibility for technology does not rest on individuals, nor on key personalities, nor even on specific institutions, but rather on a human collectivity that unfortunately lacks certainty, integrity, and solidarity regarding a common good, the meaning of personhood, and the value of community.[2]

Such a new perspective, as I envision it, must include an ethical imperative to engage in a threefold task: (1) to seek objectivity in knowledge, (2) to structure and use technology for the good of the person, and thus to preserve inner freedom before the utilization of things, (3) to remain open to a transcendental order beyond mere appearances. In short, the new perspective must include—however difficult its pursuit might seem at a time of pluralism, secularity, and liberalism—a new wisdom beyond mere rules of prudential calculus of risk and benefit.

But the scientization and technologization of politics entails the depoliticization of the masses. No dialogue or discussion aimed at a rational popular consensus can be promoted. To lead, to coordinate, to arbitrate, to show initiative or, in a word, to govern is not, however, merely or even primarily a technique of moderating debates, of participating in committee work, or of making decisions. Politics as the art of governing entails decision making by way of promoting national goals and priorities, while encouraging the public to inform itself and present its views. Politicians in a participative democracy are not mere public administrators or expert planners who guide without evaluating the meaning and validity of the goals. Politicians are not to be exclusively preoccupied with strategies or with making the system work efficiently. They must engage in reflective understanding of the proper ends of society and promote discourse with representatives of the public regarding conceptions of the common good. The right of politics to be the power that is required as a means for the implementation and coordination of matters of significance and value for the whole public is in question.[3]

Objectivity entails attention to the inner order in nature, but it also entails respect for the ontic finality that links *things* to human welfare. Objectivity demands the overcoming of disorder by the implementation of a newly designed order in its place. The power of nature over the human can and must, however, be countered by a spiritual penetration, and ordering, of the material in the service of human values, not only by an objective ordering.

This spiritual penetration demands that attention be paid to openness to the symbolism, to the beauty, and to the interconnectedness of things as they form a *system,* of which humankind might be the conscious and free agent, but never a part isolated from or indifferent to the whole. One is thus open to the mystery of things within an ordered whole or totality, which might point to a beginning and an end transcendent of human designs and understandings. Respect and reverence toward reality is a wise disposition that fosters care and conservation, and which inspires renunciation of our own power and domination. This renunciation, which is not repression but wisdom, is a credit to the new person who in liberty distances herself from an outer world in order to reach the inner order of authentic self-realization.

I would like to single out what I view as the particular role of woman in the cultivation of the wisdom recommended for the new person, entrusted with a new perspective from which to meet the challenges of the new scientific-technological culture. First of all, I would like to clarify my use of the term *woman.* I am more interested in describing a cluster of qualities that can, indeed must, exist in every human individual, than in describing women per se. These qualities as a whole I designate as the *feminine principle,* which, primarily manifested and sensed as an essential responsibility of the woman, must be present as well in the male individual. I view the human person as a dynamic interplay of complementary principles—feminine and masculine. Here I only describe the feminine principle, or the cluster of qualities descriptive of the principle that I believe are essential ingredients of the wisdom that the techno-scientific world urgently needs.

Following Carol Gilligan and Sara Ruddick, I believe that the feminine qualities inherent in personhood offer a moral sensibility and a maternal thinking, which in securing care, nurturance, and preservation serves as a counterbalance to the masculine domination, control, and efficiency associated with an emphasis on techno-science.[4] Moral sensibility with maternal thinking presupposes the qualities of humility and acceptance of the reality of conflicting interest, the recognition of the priority of holding over acquiring, persistent good humor, respect for persons and for the growth process, and flexibility in adapting to change. That is, feminization of moral thought can contribute an ingredient of crucial importance—a vital, personal, affective aspect—to an ethics of technological responsibility. The moral sensitivity of the feminine becomes a cultural value presented to our generation as decisive for moral strengthening.

Another interesting observation about women's thinking depicts her as less alienated from bodily existence than men are, and hence less likely to be seduced by abstraction, or by other forms of idealism that too easily overwhelm the masculine mind. As already pointed out, the affective component, or moral judgment, is less suppressed in woman, so her heart (gemüt) can vibrate in consonance with her convictions, and values as such are incarnated in an action that involves the entire person.

A feminine sensibility, informed by its rootedness in the concrete, emphasizes the inseparability of theory and experience or practice in the moral order. Women's moral judgments are more closely tied to feelings of empathy and compassion than those of men, and in so far as this heightened affectivity can make women more reliable, more integral, more fit for public responsibility, their education in an ethics of responsibility for the new person in the new technology dominated community becomes an urgent task for any spiritual-moral-cultural-pedagogical leader.

These qualities promote a recognition of the priority of holding and caring over conquest and acquisition. Woman in her particular respect and concern for persons and their growth promotes mutual understanding and increases her own reflective self-understanding. The feminine qualities entail a rootedness in the concrete, a discernment of how theory enlightens the practical, while practice vitalizes the theoretical.

In conclusion, the wisdom I propose is not one for a time of crisis such as suggested by Hans Jonas and Leon Kass, that is, for a time lacking a sense of being, lacking a sense of the true and the good, which then leads to moderation, prudence, and cautious humility. Their type of wisdom seems to be one that censures technology rather than one that inspires us to new perspectives; one that places barriers in front of, rather than opening up new paths for technology; one that forbids, more than commands, what must be done. I would propose, rather, a search for a wisdom that *provides* a sense of being, of good, of truth. A renewed conception of the person promises a way toward an appraisal of the primary form of being that demands our care and sense of responsibility.

NOTES

1. Hans Jonas, *The Imperative of Responsibility* (Chicago: The University of Chicago Press, 1983), 6–7.

2. On this subject one could consult Leon Kass, *Toward a More Natural* (New York: The Free Press, 1985), 5–7.

3. Joseph Agassi, *Technology* (Boston: D. Reidel Publishing Company, 1985), 250–59; and Kai Nielsen, "Technology and Ideology," in *Research in Philosophy and Technology* (Greenwich, Conn.: JAI Press, 1978), 1:135.

4. Carol Gilligan, *In a Different Voice* (Cambridge: Harvard University Press, 1982), 352; and Sara Ruddick, "Maternal Thinking," *Feminist Studies* (Summer 1980).

Deep Ecology's Mode of "Technology Assessment"

MICHAEL E. ZIMMERMAN

During the glory years of modern technology, in the decades immediately following World War II, little attention was paid to the environmental consequences of industrial development in either capitalist or socialist societies. Beginning in the United States in the early 1960s, however, especially in connection with the publication of Rachel Carson's groundbreaking book *Silent Spring,* some people began warning of the potentially catastrophic environmental consequences of rampant industrialization.[1] Despite the scientific expertise of critics like Carson, many people tended to regard environmentalists with skepticism, especially when they became associated with the countercultural movement of the day. Anything that hippies took seriously, in other words, ought not to influence the world's commitment to the conventional wisdom's long-standing goal: increasing world economic growth through global industrialization.

While the inertia of the industrial economic system was great, and while most people living in that system depended upon it for their economic well-being, many mainstream Americans became increasingly concerned about environmental issues during the late 1960s, so much so that they persuaded politicians to enact significant environmental legislation in the early 1970s, legislation that was to be enforced by the Environmental Protection Agency (EPA). One of the EPA's major tasks was to evaluate the cost-benefit analyses of the Environmental Impact Statement (EIS) now required of all public and private projects. For the first time, developers not only would have to show to what extent the proposed project would damage the environment (for example, by emitting pollutants into the air or water), but they would also have to demonstrate that the benefits of the proposed project were so significant that they would offset the environmental damage projected by the EIS.

With the institution of the EIS, *technology assessment* came to mean something very different than previously. At one time, technology assessment was a matter of evaluating the efficiency and adequacy of existing or new technology. In the dawning environmental age, a new evaluative principle was added on to technology assessment: the extent to which the existing or proposed technology damaged the environment. In 1971, the U.S. Senate utilized this evaluative principle when it turned down funding to develop the Supersonic Transport (SST). This new technology, environmentalists argued, would damage the ozone layer that protects organisms from harmful solar radiation. Many senators concluded that being able to get to Paris from New York in three hours did not justify the costs or the hazards posed by the projected new technology.

The legal provisions that resulted from the environmental reform movement in the United States and in other countries were certainly a step in the right direction. The extent to which these reforms have been effective can be measured in part by comparing American and western European environmental problems to those recently documented in eastern European countries and the Soviet Union, where environmental concerns were completely disregarded until only recently. It has been reported, for example, that half of Poland's rivers are so polluted that they are not usable even for industrial purposes. The environmental nightmares created by socialist industrialism at first make capitalist environmental problems pale by comparison. In fact, protests against the terrible pollution of land, water, and air by state industries were apparently a major factor in the downfall of socialist regimes in eastern Europe.

Despite the fact that Western nations have provided at least limited protection for the environment, governmental decrees and legislation have not been as effective as people hoped in combating air and water pollution. Alleged conflicts between "environment" and "economic growth" effectively blocked for more than a decade the passage of a new Clean Air Act in the United States. Moreover, potentially devastating environmental problems not comprehended in the early 1970s, such as global warming and deforestation, are not yet even addressed by current legislation.

In the light of the limited success of reform environmentalism, an increasing number of people are arguing that it cannot deal with the problems created by headlong industrialization combined with unchecked human population growth. These critics charge that main-line environmental organizations, such as the Sierra Club and the National Resource Defense Council, are far too willing to arrive at compromise positions with big industry and the government—posi-tions that jeopardize the well-being of the ecosphere on which all life depends. According to radical environmentalists, nothing less than a wholesale transformation of Western society (socialist and capitalist) can save the living earth—and humanity itself—from potentially dire consequences. In fact, at least one group of radical environmentalists, known as "deep ecologists," have been arguing since the early 1970s that reformist methods (including *risk assess-ment* and *cost-benefit analysis*) of addressing environmental prob-lems fail, because they are operating within the self-destructive logic of the technological system. According to deep ecologists, this system, the goals of which include economic growth for its own sake and the technological domination of nature, will eventually destroy the ecosphere upon which human life depends. While ordi-nary technology assessment uses various modeling techniques to make predictions concerning the effects of specific projects or processes on the environment, the method used by deep ecologists involves critical reflection on the metaphysical presuppositions that have led Western people to construct the very industrial-tech-nological system that gives rise to chemical plants and nuclear weapons in the first place.

A noted Norwegian philosopher and mountaineer named Arne Naess coined the term *deep ecology*.[2] Two California academics, George Sessions and Bill Devall, expanded upon Naess's concept.[3] Subsequently, a number of other writers, including Alan Drengson and Neil Evernden (Canada), Warwick Fox (Australia), Dolores LaChapelle (Colorado), Theodore Roszak and Gary Snyder (Cal-ifornia), and the present author, have contributed to the literature about deep ecology.[4] Because deep ecology is an emerging move-ment that is still defining its position, providing an accounting of it to which all adherents would agree is probably impossible. Never-theless, most deep ecologists maintain that the environmental cri-sis results from Western humanity's anthropocentric, atomistic, dualistic, and exclusively utilitarian attitudes toward the natural world. According to deep ecologists, reform environmentalism can-not effectively cope with this crisis because the recommendations it makes are based on these very same attitudes. Insofar as it remains human centered, reform environmentalism merely tries to curb some of the worst problems created by the industrial system, but doesn't challenge the system itself. For instance, reform environmentalism wants to curb pollution so that human beings will not be poisoned. Deep ecologists maintain, however, that the well-being of the entire biosphere—not just of human beings—should be taken into account when we assess the consequences of our actions.

Despite their criticism of reform environmentalism, however, deep ecologists acknowledge its importance in the short term. Moreover, deep ecologists may sometimes exaggerate the difference between their own position and that of reform environmentalists. Many alleged reformers resent being called anthropocentric and regard themselves as actively involved in moving beyond the view that nature is merely raw material for human ends.[5] Deep ecologists, then, join with many other environmentalists in maintaining that preserving a richly variegated, complex ecosystem is a noble, vital endeavor and that only a basic shift away from Western anthropocentrism and toward ecocentrism will make such preservation possible.

Tracing the roots of Western anthropocentrism has been a tricky business for deep ecologists, especially since some of them are tempted to make such anthropocentrism exclusively a trait of Western people, while in fact Eastern peoples have also displayed anthropocentric attitudes that justified practices proving ruinous to the environment. Generally, deep ecologists trace anthropocentrism to the Greco-Roman tradition, especially Stoicism, to certain tendencies in the Jewish and Christian traditions, and to the humanist ideologies of Marxism and capitalism, which may be regarded in part as secularized versions of the Jewish and Christian traditions.[6] The Enlightenment confidence in the universal progress of humanity was a major factor in establishing the radical anthropocentrism or humanism of modernity. According to such anthropocentric humanism, whether in its capitalist or communist mode, nature is nothing but raw material for enhancing human power and security. Capitalists exploit nature privately; communists do so collectively. Hence, deep ecologists dismiss as naive the tendency to regard capitalism and communism as opposites. In fact, they are two sides of the same anthropocentric coin.[7]

Not all radical environmentalists agree with deep ecology's emphasis on anthropocentrism as a major factor in the environmental crisis. Social ecologists and eco-feminists have criticized deep ecologists for overlooking what are supposedly the real roots of the environmental crisis. These roots are, according to social ecologists, hierarchalism and authoritarianism.[8] Supposedly, the same impulse at work in the social domination of one group by another is also at work in the social domination of nature: colonialism and imperialism exhibit racism as well as the exploitation of nature.[9] For example, the Spanish *conquistadores* instituted a social system that reduced New World natives to the status of slaves needed to plunder rich veins of gold and silver. English settlers in North America tended to regard native Americans as dangerous heathens, savages, and uncivilized brutes who had to be exterminated. The genocidal slaughter of native Americans by European settlers continues today in Brazil and Venezuela (as exemplified by the expansionary impact of Brazilian gold miners into the territory of the Yanomami, one of the few still largely uncontacted native tribes in the Americas).

Ecofeminists argue that the sexist, patriarchal attitude of these same *conquistadores* and of virtually all modern societies is the real root of the domination of nature.[10] In other words, the same fear of and anger at women that leads men to subjugate them is also at work in the drive of men to dominate "Mother Nature." Many social ecologists and ecofeminists agree that patriarchalism is a crucial instance of the hierarchalism and authoritarianism that justify the exploitation of humanity and nature alike. While agreeing that authoritarianism, hierarchalism, sexism, and racism are important ingredients in our currently destructive attitude toward nature, deep ecologists maintain that anthropocentric humanism is a more important ingredient. Warwick Fox has argued, for example, that one can imagine a society in which sexism, racism, authoritarianism, and hierarchy are overcome, but that still regards nature as raw material for human purposes.[11]

Some deep ecologists argue that an even more important source for environmental crisis than anthropocentric humanism is the tendency of Western people to experience the world in terms of atomistic and dualistic categories. Atomism, first developed in ancient times by thinkers such as Democritus, was revitalized by Newton and others during the seventeenth-century scientific revolution. According to atomism, everything can be analyzed into separate parts that are externally related. Nature, then, can be regarded as something "out there," apart from human beings. The perception of nature as something radically separate from humanity was reinforced by Cartesian dualism, according to which humans are essentially non-natural, nonextended *res cogitans,* mind, or intellect. Nature, by way of contrast, is portrayed as an inert, meaningless, value-free totality of matter in motion. Stripped of all intrinsic value, natural objects reveal themselves as nothing more than resources to be exploited for human ends. According to Warwick Fox, however, the "central intuition" of deep ecology is the idea that there is no firm ontological divide in the field of existence. In other words, the world simply is not divided up into independently existing subjects and objects, nor is there any bifurcation in reality between human and non-human realism. Rather all entities are constituted by their relationships. To the extent that we perceive boundaries, we fall short of a deep ecological consciousness.[12]

Subject-object dualism, which some people argue is especially prevalent in masculinist consciousness, is accompanied by profound feelings of isolation, alienation, separation, and death anxiety. Such dualism, a basic factor in all anthropocentrism, Eastern or Western, leads people to view nature and their own bodies as alien and threatening to the stability of the subject or ego. Death anxiety leads individuals and entire cultures to engage in what Ken Wilber has called the "God project," the attempt to deny death by making the ego and its products (works of civilization) immortal.[13] Arguably, then, the environmental crisis becomes acute when people already engaged in the God project develop the economic systems, and the scientific and technological skills, associated with modernity. Given the powers of modern technological society, people are seduced into believing that they can defeat death by conquering nature.

While many deep ecologists appeal to trends in contemporary science as evidence that the dualistic way of thinking is being overcome, they are also critical of early modern science, which enthroned the atomistic way of seeing things. Yet if the modern scientific world view was atomistic, it was not necessarily anthropocentric; indeed, by describing humans as clever animals living on an insignificant planet in the middle of an incomprehensibly vast universe, modern science punctured the anthropocentric myths of the Middle Ages. When the modern scientific view was combined with anthropocentric humanism during the Enlightenment, however, there occurred a synergistic effect that greatly enhanced the human domination of the natural world. According to *naturalistic humanism,* humans are both: (1) clever animals competing in an ultimately meaningless struggle for survival and dominance with all other forms of life; and (2) the source of all meaning, purpose, and value. At first glance, of course, naturalism and humanism would seem to be incompatible, if not contradictory. In fact, however, nineteenth- and twentieth-century people have overlooked the conflict between these two attitudes and have successfully combined them into the following, often unspoken, principle: humans are justified in their domination of nature, not only because this is simply natural law, but also because humans are the source of all value. Hence, humans have the right to do what they want with nonhuman beings. The ratification of human rights over nature is a central dimension to both communism and capitalism, each of which is a representative of naturalistic humanism. Both of these political systems have an exclusively economic-utilitarian view of nature. Nature is simply assumed to be the property (private or collective) of humans.

Deep ecologists have argued that while reform environmentalism is important for the short term, nothing less than a transformation of human self-understanding is necessary, if humanity and the ecosphere are to survive in the long run. Such a transformation would involve replacing anthropocentrism with ecocentrism, atomism with relationalism, dualism with nondualism, and utilitarianism with an attitude of respect and love for all human and nonhuman beings alike. Deep ecologists do not believe that this new attitude can be developed by tinkering with existing Western ethical systems, for example, assigning "rights" to nonhuman beings, or in developing a doctrine of the "inherent value" of nonhuman beings.[14] For one thing, the notion of rights for animals, plants, and rivers seems incomprehensible, since rights are generally correlated with responsibilities. Moreover, the doctrine of rights seems inextricably involved with the metaphysical atomism, which deep ecology seeks to overcome.[15] Further, anthropocentric attitudes are often discernible in the ways in which we assign value to nonhuman beings. For example, animals are said to be more valuable than plants because animals are more conscious or more sentient, but consciousness or sentience is what humans prize most about themselves; hence, humans tend to value those nonhuman beings that exhibit the highest level of consciousness or sentience. As John Rodman has asked, "Is this the new enlightenment—to see nonhuman beings as imbeciles, wilderness as a human vegetable?"[16]

Aware of the problems posed by the attempts to develop a new ethical basis for human dealings with nature, deep ecologists argue that in establishing a new humanity-nature relationship, ontology must precede ethics. That is to say, a new mode of understanding what things *are* will spontaneously bring forth behavior consistent with that understanding. For example, because I identify myself with my body, I spontaneously take care of it. I do not have to follow moral precepts to avoid cutting off my fingers or burning my foot. Similarly, if I were to expand my sense of identity to include beings other than myself, I would avoid doing harm to them, too. Deep ecologists, then, call for an expanded sense of self or a broadened identification with all beings, not just with my own ego or my own body.[17] This broadened self-identification is very similar to the teachings of Buddhism and other nondualistic traditions. Nondualistic awareness discloses the self in all things; such awareness spontaneously generates an attitude of care toward all things. Warwick Fox has suggested that deep ecology be renamed "transpersonal" ecology, to reflect the fact that it calls for an all-encompassing mode of awareness that goes beyond the confines of the personal "ego."[18]

While the insight required for nondual experience is often associated with mysticism, deep ecology also appeals to the findings of contemporary science to support its contention that there is an alternative to the atomistic-dualistic way of seeing things. In physics, biology, and many other disciplines, researchers now maintain that the atomistic and mechanistic world view is no longer tenable. Instead of the universe being a machine composed of separate parts, it is much more like an organism that is internally related and still in the process of evolving. Current findings suggest that Leibnitz may have been on the right track in saying that every aspect of the universe mirrors or reflects every other aspect. When we describe our experience in dualistic terms, we say that I, the subject, am experiencing something "in here," while nature is "out there." Deep ecologists maintain, however, that it makes far more sense to say that my experience is an event of the universe itself; the universe becomes aware of itself through me and through all other dimensions of itself. Deep ecologists emphasize the importance of the fact that the universe contains within itself the capacity for generating carbon based forms of life that are capable of eating, sleeping, reproducing, swimming, and even thinking and talking.

Arne Naess uses the term *self-realization* to describe the expanded sense of self required for the deep ecological attitude. Self-realization is a term that is often associated with anthropocentric humanism, but by self-realization Naess does not have in mind the achievement of the atomized ego's goals of infinite expansion and acquisition. Following Buddhism and Spinoza, Naess defines self-realization as a transformation of the ego, as a revelation that a person's authentic destiny involves identifying not just with one's own desires, or with one's own body, or even with one's own species, but with the living cosmos, the larger "self" of which we are all manifestations.

It may sound paradoxical, but with a more lofty image of maturity in humans, the appeal to serve deep, specifically human interests is in full harmony with the norms of deep ecology. But this is evident only if we are careful to make our terminology clear. This terminology is today far from common but it may have an illuminating impact. It proclaims that essentially there is at present a sorry underestimation of the potentialities of the human species. Our species is not destined to be the scourge of the earth. If it is bound to be anything, perhaps it is to be the conscious joyful appreciator of this planet as an ever greater whole of its immense richness. This may be its "evolutionary potential" or an eradicable part of it.[19]

Naess, then, calls for a higher humanism to replace the anthropocentric humanism that is largely responsible for the technological domination of nature. In this respect, and in others as well, Naess sounds like Martin Heidegger, who said that Western humanity needs a higher humanism to move it beyond the one-dimensional, anthropocentric humanism that discloses all things to be raw material. Several authors, including myself, have attempted to elucidate the apparent congruity between Heidegger's thought and deep ecology.[20] In view of recent disclosures about the scope and duration of Heidegger's commitment to national socialism, however, I have come to the conclusion that we must question the wisdom of attempting to make Heidegger a philosophical forerunner of deep ecology. A brief examination of Heidegger's thought and its apparent proximity to aspects of deep ecology, will enable us to offer some critical remarks about deep ecology. Some critics charge that deep ecology amounts to eco-fascism and use Heidegger's support for nazism to justify their claims. While such critics overstate their case against deep ecology, we must be wary of the political implications of claims advanced by some people sympathetic to deep ecology. Deep ecologists such as Arne Naess, who was active in the resistance against the Nazi occupation of Norway, are well aware of the dangers of using talk about human "rootedness" in the earth to justify fascist social proposals.

According to Heidegger, the contemporary destruction of nature is the inevitable outcome of the history of being, the story of how Western metaphysics has unfolded from the time of Plato to the present. By "being," Heidegger meant not the absolute foundation for things (such as the Creator God), but rather the very "presencing" or "showing up" of things. For something "to be," then, means it is present, manifest, or unconcealed. For such self-manifesting to occur, a clearing or site is required. Human existence constitutes this historical-linguistic clearing. The history of being is really the history of the self-concealment of being or presencing, such that humans have focused on the entities that have manifested themselves. We are so fascinated with and dependent on entities, in other words, that we overlook the astonishing ontological event of their *presencing*. Plato and Aristotle initiated the history of metaphysics by conceiving of being not as presencing, but rather as a superior entity that is constantly present. For Plato, this constantly present entity was the eternal form *(eidos)* that was the model for empirical things. By conceiving of being as a blueprint or model, Plato began the tradition of productionist metaphysics, the tradition that holds that for something "to be" means for it "to be produced."

Ultimately, this productionist model culminated in the nihilistic era of modern technology, in which for something to be means for it to be raw material in the ever-expanding and self-enhancing process of production. In the technological era, being has completely effaced itself. All humans encounter are entities. Conceiving of themselves as clever animals struggling to survive, humans turn *logos* into mere calculating intelligence designed to dominate nature. Since human beings are themselves part of nature, modern technology also reveals them as raw material. Instead of using technology for human ends, modern humanity has become the means for the ends of technology itself. Heidegger used the term *Gestell* to describe the present age in which we are compelled to understand everything one-dimensionally as standing reserve for increasing technological control over nature. For Heidegger, things manifest themselves through language. Today, the technological language holds sway. What this age can reveal to us, Heidegger believed, is that humanity does not speak language, but rather that language speaks humanity. The possibilities for our behavior, in other words, are determined by a destiny that lies beyond our capacity to control or even to understand.[21]

Like deep ecologists, Heidegger criticized the anthropocentric, atomistic, dualistic, and utilitarian categories that determine how Western humanity understands itself and nature as well. Moreover, Heidegger also called for a radical transformation of human existence, a transformation that would free us from the compulsion to control everything and permit us to "let things be." This transformed condition, *Gelassenheit,* seems very close to what Naess and other deep ecologists have in mind by self-realization. Heidegger's authentic self is not an ego-subject bent on dominating things, but is rather the clearing owned by being itself. The authentic self is the historical-linguistic site through which entities may manifest themselves in ways other than as raw material. Presumably, an authentic humanity would dwell on earth in a way that would "let things be."

While they do not make the distinction between being and entities (the "ontological difference"), which is so central to Heidegger's thought, deep ecologists have come to regard Heidegger as a crucial example of the minority tradition in Western thought that calls for an alternative to the dominant attitude of anthropocentric humanism. Although sympathetic both to deep ecology and to Heidegger's thought, I believe that there are two important reasons why deep ecologists should adopt a critical stance toward it. The first reason is Heidegger's notorious involvement with national socialism; the second is his antinaturalistic attitude.

The question of Heidegger's political affiliation was revived in 1987 in Paris by the publication of Victor Farias's book, which was indebted to the work of the German economic historian Hugo Ott.[22] Farias and Ott have demonstrated that Heidegger's support for national socialism was far more profound and enduring than Heidegger himself claimed in his postwar defense of his Nazi past. Heidegger's Nazi speeches in 1933–34, as well as his lectures and letters during the 1930s and early 1940s, indicate that he regarded his own thinking, and the poetry of Hölderlin, to be the spiritual guideposts for the national socialist revolution. Indeed, Heidegger attempted to make himself Germany's spiritual *Führer*. Heidegger believed that Hitler's revolution would save Germany from the fate of industrialism and modernity, which had already destroyed Russia and America. Hitler had come to power on a platform that rejected both communism and capitalism, and that depicted the Enlightenment as the worst thing ever to befall humanity. Enlightenment rationalism removed the mystery from things, turned nature into a mere resource, and transformed humans into calculating intellects or exploited laborers. Appealing to a romanticized German past of small towns, farms, and countryside, Hitler and many of his colleagues conjured up visions of a nature mysticism in which Germans would recover their sense of relatedness to a natural world threatened by modern technology. Hitler's rhetoric found a ready audience in a population that had experienced the often ghastly and alienating effects of industrialization. With his talk of the need for the German *Volk* to regain its "rootedness" in the soil, however, Hitler blended his infamous racist doctrines that argued that Jews and other non-Aryan peoples were polluting the blood of the German *Volk* and would thus have to be removed (forcibly, if necessary) from Germany. Because of the admixture of such racism with the "green" dimension of Nazi politics, the postwar environmental movement in Germany could only arise in the late 1960s, and then only under the guise of leftist politics.

That Heidegger used his thought to support national socialism, and that his thought appears to be in some ways consistent with deep ecology, is a cause for concern on the part of critics who fear that deep ecology is promoting a kind of eco-fascism.[23] This concern is justified in certain respects. First of all, deep ecologists often accept rather uncritically the Heideggerean account of Western history as a decline from more pristine beginnings. The Enlightenment, according to this account, was simply a late stage in the history of a power-tripping Western humanity bent on dominating nature. While modern Western history does include a drive to control natural forces, modernity cannot fairly be described in the wholly negative way that we sometimes encounter in the writings of deep ecologists. The view of modernity as the culmination of a long process that began when humanity tore itself away from its "appropriate" relation to the natural world is consistent with the reactionary view of human history promoted by many Germans of Heidegger's generation. Denying that there has been any evolutionary social and political development in Western history, it fails to see that there was an authentic emancipatory dimension to the Enlightenment, despite the limited character of the Enlightenment's scientific and political vision. By adopting uncritically Heidegger's world view and that of his contemporaries, people end up with the reactionary view that the whole of modernity—including democracy, liberalism, rationalism, science, modern technology, capitalism, and communism—is corrupt. In addition to its indiscriminate lumping of different political and cultural trends, such an interpretation fails to appreciate the importance of the political project of modernity: to use reason to free people from authoritarian regimes and to found democratic societies emphasizing political liberty and economic justice.

When the call by deep ecologists to overcome anthropocentrism is mixed with what amounts to a reactionary view of Western history, one can understand why some critics regard deep ecologists with suspicion. Such suspicion was fueled some time ago when people affiliated with the deep ecological organization Earth First! made misanthropic and apparently racist remarks and proposed draconian population control measures. Despite such statements on the part of political activists associated with deep ecology, no one will find racist or protofascist remarks in the writings of deep ecologists. Indeed, they call for social developments—such as decentralization and bioregionalism—that would undermine the totalizing impulse at work in the political and economic institutions of modernity. Nevertheless, where deep ecologists must be cautious, and where Heidegger failed, is to be sure that their critique of anthropocentric humanism and metaphysical atomism does not undermine one of the major achievements of Enlightenment modernity, namely, the notion that individuals have the right both to form democratic institutions and social relations, on the one hand, and to protect themselves from authoritarian, irrational, and racist regimes that would sacrifice individuals and their freedom for the sake of a "higher good," on the other.

In my view, a necessary condition for a shift to a nondualistic, ecocentric mode of awareness is the global consolidation of the view that all individuals (no matter of what class, race, creed, or ethnic background) are worthy of respect and are endowed with certain inalienable rights. Such a view is consistent with the contemporary call for "celebrating cultural difference," but also makes clear that such a celebration presupposes adherence to the belief that all humans—regardless of specific cultural or racial origins—are worthy of respect and tolerance. Without such a belief, difference is not celebrated but instead made absolute, in such a way that the "other" may become regarded as nonhuman. Because many Germans rejected the Enlightenment commitment to toleration and to universal humanity, they asserted the uniqueness and superiority of the German *Volk*—and thus justified the exclusion and removal of non-Germans from their midst. Moreover, because many Germans had not sufficiently internalized the importance of individual responsibility, they were willing to regress to an earlier, collectivist stage of awareness in which they sacrificed their individual responsibility to an allegedly higher authority. As environmental problems grow worse in the future, we must be alert to the possibility of the emergence of groups promoting something like eco-fascism.

Only as respect for human persons grows can we expect to see the demand grow for respectful treatment of the living earth itself, for we will see ever more clearly that destroying the earth undermines the conditions necessary for healthy human life. It may be that as we increasingly perceive the relationship between healthy, flourishing people and a healthy, flourishing planet, anthropocentrism will gradually give way to ecocentrism. In the light of our growing comprehension of the interrelationship and interdependence of all people and all things, we may learn to extend respect and care to the whole living earth. While important, such extension would not yet constitute overcoming the dualism that currently characterizes humanity's relationship to the natural order, for overcoming such dualism will require centuries of continuing human evolution. In the meantime, it is important to sustain the political, social, and cultural conditions necessary for this step in human evolution. Above all, we must be clear that attempts to overcome humanity-nature dualism by a regression to earlier collectivist stages of human awareness have proved and will continue to prove catastrophic.

The second problem involved with linking Heidegger and deep ecology is that Heidegger had a profoundly antinaturalistic attitude, while deep ecology emphasizes the internal relation of humanity with the rest of nature. Heidegger refused to accept the Darwinian notion that humans were descendants of apes. Indeed, in rejecting Nietzsche's claim that humans are "clever animals," he defied the metaphysical tradition begun by Aristotle's definition of the human as a "rational animal." Heidegger maintained that this conception of humanity culminated in the technological age, in which the clever animal competes against all other beings for survival, control, and power.

In his lectures on Hölderlin, Heidegger attempted to develop an alternative to the Western conception of humanity and nature. Instead of viewing humans as clever animals who use language to dominate everything, Heidegger encouraged the view that humans are the vessel through which language speaks, that is, the opening through which entities are manifested through language. Rejecting modern science's view of nature as the totality of matter-energy operating according to strict causal necessity, Heidegger portrayed nature as *physis,* the event of presencing by which entities bring themselves forth into manifestness. *Physis* (or being) lays hold of human existence as the site through which *physis*/presencing can occur. Two ways in which entities can present or manifest themselves are as objects for scientific study or as raw material for industrial usage. While currently in the grip of such one-dimensional ways of disclosing entities, Heidegger maintained that a radical reversal was possible in which entities could display themselves in alternative ways—ways that were somehow richer and more variegated.

In emphasizing that the event of presencing is more fundamental than the entities that are present, and in stressing that human existence is the nothingness or clearing in which such presencing occurs, Heidegger verges on a kind of ontological phenomenalism that calls to mind Mahayana Buddhism. The deep ecologist Arne

Naess has also emphasized the extent to which self-realization involves the direct apprehension of the fact that "things" are in fact patterns or gestalts that have no "self-nature," but are rather phenomena in a cosmic play.[24] Like Heidegger, then, Naess questions the adequacy of and seeks an alternative to the materialistic, mechanistic view of nature. Unlike Heidegger, however, Naess is far more open to the possibility that the findings of contemporary science will provide insight conducive to the deep ecological attitude.

In conclusion, let me reiterate that deep ecology provides a radically different methodology to risk assessment in the face of modern technology. For deep ecologists, the risk posed by modern technology is planetary in scope; hence, nothing less than a profound critical reflection on our world view is required if we are to avert global disaster. In calling for an end to anthropocentric humanism, metaphysical atomism, dualism, and resource utilitarianism, deep ecologists are reminiscent of Heidegger's critique of modernity. Both deep ecologists and Heidegger call upon us to let things be, but Heidegger's relation to national socialism rightly gives us pause when we think of his conceptual proximity to deep ecology.

Deep ecology differentiates itself from Heidegger, however, in two respects. First of all, deep ecologists generally adopt the liberatory language and hope of the Enlightenment, despite their critique of Enlightenment anthropocentrism. The Enlightenment postulated the notion of human progress. Without a commitment to the possibility that humans can evolve beyond the limits of anthropocentric humanism, deep ecologists would have no reason to expect that humanity could ever move beyond its current dualistic attitudes. Second, deep ecologists also take more seriously than did Heidegger the possibility that modern (or postmodern) science might reveal something true about human existence.[25] Hence, while deep ecologists criticize modernity, they are far less likely than Heidegger to reject out of hand its political achievements and to adopt a reactionary social attitude that would be disastrous if it were resurrected in today's world. Deep ecologists have the advantage of being able to learn from the mistakes made in the first part of this century. They recognize that while they may continue to harbor utopian hope for a radical transformation of the humanity-nature relationship, they must nevertheless face up to the dangers posed by the politics of redemption.

In some sense, deep ecology provides a metaphysical critique of the shortsightedness of that mode of technology assessment that fails to appreciate the metaphysical origins and ecological dangers posed by a heedless modern technology. Deep ecology's critical assessment of the origins and limitations of modern technology, however, must always be accompanied by a penetrating evaluation of its own critical methods and political presuppositions. Only in this way can deep ecologists continue to discover their own limitations as well as their strengths. The resolution of complex cultural and environmental problems takes much time, patience, and the dialogical participation of people of many different persuasions. Deep ecology is certainly one of the important voices in the crucial dialogue now taking place regarding the future of humanity and the earth.

NOTES

1. Rachel Carson, *Silent Spring* (New York: Houghton Mifflin, 1962). Then secretary of the interior Stewart Udall's book, *The Quiet Crisis* (New York: Holt, Rinehart & Winston, 1963) also made an important contribution to awakening people to environmental problems.
2. Arne Naess, "The Shallow and the Deep, Long-Range Ecology Movement: A Summary," *Inquiry* 16 (1973): 95–100. For a fuller exposition of Naess's views, see his book *Ecology, Community, and Lifestyle: Outline of an Ecosophy,* translated and revised by David Rothenberg (New York: Cambridge University Press, 1989).
3. Bill Devall and George Sessions, *Deep Ecology: Living as If Nature Mattered* (Layton, Utah: Gibbs M. Smith, Inc., 1985).
4. Alen Drengson, *Shifting Paradigms: From Technocratic to Planetary Person* (Victoria, B.C.: LightStar Press, 1983); see also Drengson's journal of "ecosophy," *The Trumpeter* (Victoria, B.C.: LightStar Press); Neil Evernden, *The Natural Alien* (Toronto: University of Toronto Press, 1985); Warwick Fox, *Toward a Transpersonal Ecology: Developing New Foundations for Environmentalism* (Boston: Shambhala, 1990); Dolores LaChapelle, *Earth Wisdom* (Los Angeles: Guild of Tutors Press, 1978), and *Sacred Land, Sacred Sex, Rapture of the Deep: Concerning Deep Ecology—and Celebrating Life* (Silverton, Colorado: Finn Hill Arts Press, 1988); Theodore Roszak, *Person/Planet: The Creative Disintegration of Industrial Society* (Garden City, N.Y.: Anchor/Doubleday, 1978); Gary Snyder, *Turtle Island* (New York: New Directions, 1974), and *The Real Work* (New York: New Directions, 1980); Michael E. Zimmerman, "Toward a Heideggerean Ethos for Radical Environmentalism," *Environmental Ethics* 5 (Summer 1983): 99–131; Zimmerman, "Implications of Heidegger's Thought for Deep Ecology," *The Modern Schoolman* 54 (November 1986): 19–43.

For an outstanding bibliographical review of deep ecology, see George Sessions, "Shallow and Deep Ecology: A Review of the Philosophical Literature," *Ecological Consciousness: Essays from the Earth Day X Colloquium,* ed. Robert G. Schultz and J. Donald Hughes (Washington, D.C.: University Press of America, 1981).
5. In reviewing Bill Devall's *Simple in Means, Rich in Ends: Practicing Deep*

Ecology, for example, the executive director of the Sierra Club, Michael Mc-Closkey, strongly disagreed with Devall's criticism of so-called reform environmentalism: "Basically, [Devall] believes that deep ecologists are biocentric and that reform environmentalists are anthropocentric. Here Devall is just plain wrong. It is the developers who champion the Dominant Social Paradigm, and environmentalists of almost all stripes who champion the New Ecological Paradigm. There are differences among environmentalists in interpreting and applying this new paradigm, but it is grossly misleading to suggest that reform groups have the same views as developers, or that they are basically anthropocentric." *Sierra* 74 (January/February 1989): 162–64.

6. On this topic, see John Passmore, *Man's Responsibility for Nature* (New York: Charles Scribner's Sons, 1974). In a famous essay, "The Religious Roots of Our Ecologic Crisis," *Science* 155 (19 March 1967), Lynn White, jr., argued that Christianity was largely responsible for Western humanity's arrogant and destructive treatment of nature. For a critical appraisal of this view, see the essays referred to in the bibliography of Carl Mitcham and Jim Grote's collection, *Theology and Technology: Essays in Christian Analysis and Exegesis* (Lanham, Maryland: University Press of America, 1984). In *The Coming of the Cosmic Christ* (New York: Harper & Row, 1988), Matthew Fox makes a strong case for what amounts to a deep ecological reinterpretation of Christianity by emphasizing its mystical rather than its redemption tradition.

7. On this topic, see my essay "Anthropocentric Humanism and the Arms Race," in *Nuclear War: Philosophical Perspectives,* ed. Michael Fox and Leo Groarke (New York: Peter Lang, 1985).

8. See Murray Bookchin, *The Ecology of Freedom* (Palo Alto: Cheshire Books, 1982).

9. For a critical examination of the social ecology of Murray Bookchin, see Robyn Eckersley, "Divining Evolution: The Ecological Ethics of Murray Bookchin," *Environmental Ethics* 11 (Summer 1989): 99–116. Bookchin replies in "Recovering Evolution: A Reply to Eckersley and Fox," *Environmental Ethics* 12 (Fall 1990), 253–74.

10. For a helpful analysis of ecofeminism, see the first footnote of Karen Warren's essay, "The Power and the Promise of Ecological Feminism," *Environmental Ethics* 12 (Summer 1990): 125–46. The first footnote of Warren's provides a good bibliography of essays in ecofeminism. Irene Diamond and Gloria Feman Orenstein have recently published an excellent collection of essays on ecofeminism: *Reweaving the World: The Emergency of Ecofeminism* (San Francisco: Sierra Books, 1990). For a sympathetic but critical review of ecofeminism, see my essay, "Feminism, Deep Ecology, and Environmental Ethics," *Environmental Ethics* 9 (Spring 1987): 21–44.

11. Warwick Fox, "The Deep Ecology-Ecofeminism Debate and its Parallels," *Environmental Ethics* 11 (Spring 1989), 5–26.

12. See Warwick Fox, "Deep Ecology: A New Philosophy of our Time?" *The Ecologist* 14, no. 5/6 (1984): 194–200; quotation is from p. 196.

13. See Ken Wilber, *Up From Eden: A Transpersonal View of Human Evolution* (Boulder: Shambhala, 1981), and several other of Wilber's insightful books.

14. The best critique of the axiological approach to the humanity-nature relation is found in Fox, *Toward a Transpersonal Ecology.*

15. On this topic, see my essay, "The Crisis of Natural Rights and the Search for a Non-Anthropocentric Basis for Moral Behavior," *The Journal of Value Inquiry* 19 (1985): 43–53.

16. John Rodman, "The Liberation of Nature?" *Inquiry* 20 (1977): 83–131; citation is from p. 94.

17. See Arne Naess, "Self-Realization in Mixed Communities of Humans, Bears, Sheep, and Wolves," *Inquiry* 22 (1979): 231–241; Naess, "A Defence of the Deep Ecology Movement," *Environmental Ethics* 6 (1984): 265–70; Naess, "Identification as a Source of Deep Ecological Attitudes," in *Deep Ecology,* ed. Michael Tobias (San Diego: Avant Books, 1985); Fox, *Toward a Transpersonal Ecology.*

18. Fox, *Toward a Transpersonal Ecology.*

19. Arne Naess, "The Arrogance of Anti-Humanism?" *Ecophilosophy* 6 (1984): 9.

20. See my essays on Heidegger and deep ecology listed in footnote three, as well as Bruce V. Foltz, "On Heidegger and the Interpretation of Environmental Crisis," *Environmental Ethics* 6 (Winter 1984): 323–38; Laura Westra, "Let It Be: Heidegger and Future Generations," *Environmental Ethics* 7 (Winter 1985): 341–50; Dolores LaChapelle, *Earth Wisdom,* 80–86, 92–96.

21. For a more detailed account of Heidegger's interpretation of the origins and character of modern technology, see my book *Heidegger's Confrontation with Modernity: Technology, Politics, Art* (Bloomington: Indiana University Press, 1990).

22. Victor Farias, *Heidegger et le nazisme* (Paris: Verdier, 1987). Joseph Margolis and Tom Rockmore laboriously edited what amounts to a definitive edition of this book that takes into account changes made when Farias published the book in Germany. See *Heidegger and Nazism,* trans. Paul Burrell, with advice of Dominic Di Bernardi (French material) and Gabriel R. Ricci (German material) (Philadelphia: Temple University Press, 1989). Most of the important material that Farias utilized from Ott's work is found in Ott's book, *Martin Heidegger: Unterwegs zu seiner Biographie* (Frankfurt am Main: Campus, 1988). I have examined the Heidegger-Nazi controversy in several places, including *Heidegger's Confrontation with Modernity;* "L'Affaire Heidegger," *Times Literary Supplement* no. 4,462 (7–13 October 1988): 1115–17; and "The Thorn in Heidegger's Side: The Question of National Socialism," *The Philosophical Forum* 20 (Summer 1989), 326–65.

23. The most noted and vociferous critic of deep ecology's fascist potential is Murray Bookchin.

24. See Arne Naess, "The World of Concrete Contents," *Inquiry* 28 (December 1985): 417–28.

25. In *The Reenchantment of Science* (Albany: SUNY Press, 1988), David Ray Griffin has gathered an outstanding collection of essays that show the extent to which contemporary scientific research calls into question the dualistic, mechanistic, and often anthropocentric attitudes of classical modern science. The essays show quite clearly that these changes in our scientific understanding of the nature have profound implications for our own self-understanding and for our conception of humanity's place in the natural order.

Science, Technology, and the Military

CARL MITCHAM

One exoteric function of philosophy is to reflect on public problems in an attempt to identify those contemporary responses that deserve, from the most general perspective, our considered support. The future at issue need not connote unambiguous progress, only a helpful readjustment in the contingencies of human affairs. There is also a place for the criticism of less-than-adequate ideas associated with such historical tendencies. While the owl of Minerva takes wing at dusk, the cock of Athena crows at dawn— occasionally to warn of hidden dangers.

What follows, then, is an exercise not in the flight of speculation but in more humble, although nonetheless worthy, tasks—with regard to the relationship between science, technology, and the military. This effort is, however, quite preliminary and grows out of involvement with a larger project on "Ethical Issues Associated with Scientific and Technological Research for the Military."[1] It is offered here because of a sense that one of the important "new issues" defined by the conquest of "new worlds" through "new technologies" is precisely the engagement of science and technology with the military.

TWO VERSIONS OF THE PROBLEM OF SCIENCE, TECHNOLOGY, AND THE MILITARY

In public discourse of both the old world and the new, it is not uncommon for a link between science, technology, and the military to be acknowledged—and decried. This has not always been the case. But today there is a sense of uneasiness among members of the scientific and technical community about association with the military. Even those who argue the social exigency of the relationship often admit that military priorities may well distort science and technology. (Within the military community, too, there are reservations, perhaps to the effect that science and technology today distort the military by substituting high-tech weapons that do not always work for more basic military hardware and discipline.) Outside the scientific and technical community there is an even stronger sense that institutional associations aggravate the worst aspects of each.

In the first, or negative, statement of the problem, it is suggested that there are necessary but destructive links between science, technology, and the military. Clearly science and technology find their destructive aspects emphasized by the military and, correspondingly, military destructiveness is excessively enlarged by science and technology. Indeed, from this perspective, there is such a *prima facia* case for the thesis that militarism is an existentially necessary (if not always sufficient) condition for intensive technological development that the burden of proof should rest on those who would reject it—although in fact it is the other way around, and this idea remains a radical minority view. Such a view nevertheless presents a major challenge to dreams of reformed, humane, alternative, or ecological technologies—themselves necessities for our continued existence.

The challenge exists even when it is admitted, as it must be, that technology is not exclusively wedded to warfare, but also exhibits polygamous involvements with peace. Although technology engenders both good and evil possibilities, the mere accidental reduction of any catastrophic risk to destructive actuality could transform all previous golden achievements into dead-end illusions. The military connection is the concrete manifestation of such fateful potential in the midst of the bright attractions of science and technology.

The common presumption of all responses to radical criticisms of a "military-industrial complex" is that the nexus is accidental or contingent: it need not be the way it is. We can have our technological advances and dynamic technological culture in a demilitarized form. Militarism is just an illness that can be cured. If it does not willingly lie down on the couch of analysis, it can at least be constrained by the therapy of democratic social control. However, that this complex is something less than the simple result of an unfortunate childhood is suggested by more than one consideration. First, warfare is properly divided into historical periods precisely on the basis of its distinctive technologies. There is Stone Age warfare, Bronze Age warfare, Iron Age warfare. Modern military history is ushered in by the use of gunpowder and the printing press (to disseminate nationalist propaganda). Contemporary military history begins with the introduction of radar, jet aircraft, and nuclear weapons.[2]

Furthermore, the development of warfare waged by standing armies only arises with the technological creation of "unnatural" or exceptional wealth by the primary great civilizations in China, India, Mesopotamia, and Egypt.[3] With the creation of exceptional wealth there results the need to defend it against those who "naturally" desire to share it. The decisive move in Plato's *Republic* from the "true" and "healthy" city, which needs no armed forces, to the "feverish" one, with its large contingent of guardians, turns precisely on this issue of technological change.[4] At the same time, warfare and technical change mutually stimulate one another. Thucydides, for instance, has Corinthians predict that Spartan resistance to technical change may lead to defeat by the more technically aggressive Athenians in the Peloponnesian War.[5]

Third, both science and technology are often associated with military images. One need only recall the facile references to "wars" against polio, small pox, cancer, hunger, or AIDS. Indeed, it has been persuasively argued that only when nature is imagined as something to be conquered do modern science and technology become viable social institutions.[6] Certainly, many contemporary supporters of sustained scientific and technological prowess argue—usually from behind the scenes—that only fear of war can galvanize a democratic public to contribute with sufficient generosity to create a world-class scientific establishment.[7] Even more explicit is the optimistic theory, advanced by philosophers of the Enlightenment, of industry and commerce as effective substitutes for human warfare. For the Encyclopedists, warfare of human beings against each other is caused by the competitive pursuit of scarce material resources and surplus human energies directed toward the achievement of recognition or glory. Scarcity can be overcome by the technological conquest of nature, which not only absorbs human energy but can provide its own special form of honor and glory. In the essays of David Hume, technology is already the "moral equivalent of war" for which William James will later seek.[8]

A pessimistic revision of this Enlightenment view—originally espoused by the nineteenth-century inventor, Alfred Nobel, and later at length by the military historian Jan Bloch—is that increasingly advanced weapons can render human warfare unthinkable or at least controllable.[9] This is, of course, the basis of nuclear deterrence. But the underside of this virtuous repression promoted by advanced technological weapons is a "return of the repressed" in a more intensive technological assault upon nature through megaconsumer waste dumps and atmospheric ozone depletion, not to mention the colossal economic loss of weapons development and manufacture.

Finally, metric analyses of technological progress (i.e., numbers of engineers, patents) reveal modernity as shot through with exponential increases. Of all the inventions in the history of the world, more than half have been made in our own lifetimes. But a metric analysis of modern warfare in terms of numbers of wars, casualties, etc., reveals a similar curve. Of all the destruction that has taken place in warfare throughout history, surely a majority of it has occurred also during the last century.[10]

The reform of technology is a well-recognized necessity of our time. From technologies of war must be created technologies of peace; from technologies of pollution, technologies in harmony with the environment; from technologies of domination, technologies of conviviality. Human life is not possible without technology.

But the bonds forged especially between modern technology and war are not easily broken—especially when militarism is recognized as exhibiting metaphorical as well as literal powers. The association of science, technology, and the military, in the new world and the old, is a major factor in the environmental, social, and cultural disruptions commonly attributed to scientific and technological change. But if such bonds are to be broken only by going to war against them, could success not entail a kind of failure? Should not the glibness with which even pacifist William James refers to "the war against war" at least give pause?[11]

In the second or positive statement of the problem, the emphasis is on inner tensions between science, technology, and the military. Like the negative argument, the positive version of the problem acknowledges, even emphasizes, the corrupting influences of the military on science and technology. But the idea that the military has corrupted science and technology presupposes that science and technology could have a reality independent of and uncontaminated by military imperatives or influences. This is thus a version of the problem popular within the technical community, although one strongly reflected in public beliefs and rhetoric as well.

According to the positive statement of the problem there is an inherent tension between the ways of life associated with the three institutions of science, technology, and the military. The scientist must open eyes and ears to all that manifests itself; he or she must keep hands off, only touch with the eyes. But the technologist closes the eyes, not to be restricted by what simply is, and instead lets the imagination direct the hands to reach out and construct a world. The technologist escapes from what science knows, nature, into what engineering constructs, artifice. The soldier, like the technologist, must put hands on; but, unlike the technologist, he or she must de-construct, putting hands on artifice to break it down and return it to nature.

More concretely, scientists favor free flow of information and want to publish their results in order to gain prestige and recognition. Engineers may not want to publish results, but they certainly want to patent them. But the military wants to restrict the processes of both publishing and patenting, lest a rival military power gain access to its own costly information and inventions. The real-world linkages of science, technology, and military in the democratic state can be said to resolve these tensions via trade-offs between competing professional interests—and at the expense of failing to consider what a human being qua human should do.

PHILOSOPHICAL ANALYSES OF THE RELATIONSHIP BETWEEN SCIENCE, TECHNOLOGY, AND THE MILITARY

Theoretical articulations of the problem of the science-technology-military relationship are adumbrated in the final works of two major twentieth-century philosophers in the period between the great wars of the first half of the twentieth century. Both Henri Bergson's *The Two Sources of Morality and Religion* (1932) and Edmund Husserl's *The Crisis of European Sciences and Transcendental Phenomenology* (1934–38) postulate a deformed growth of science and technology in relation to the military. Bergson's work elucidates what has been called the negative version of the problem, Husserl's the positive. What is remarkable is how each further proposes the correction of deformity through a reappropriation of origins, that is, through an appeal to life.

For Bergson there exists a fundamental unity between science-technology and warfare. Human life has been so constituted that war is "inevitable"[12] and "natural."[13] By endowing humanity with "fabricating intelligence" instead of specific instruments, nature sets the stage for conflict.[14] Because human beings make instruments or tools, tools are things that they own, property—and, thus, things that can be taken away.

The origin of war is property . . . and since humanity is predestined to property by its structure, war is natural. The war instinct is so strong that it is the first to appear when one scratches away civilization in search of nature.[15]

The natural impetus toward self-preservation that serves as the foundation of human creativity in the form of tools and weapons (and their devious uses) in both craft and warfare further drives human beings to form close-knit, compact, hierarchal societies. Like Hobbes before him, Bergson sees vital exigencies at work in the creation of authoritarian states and their closed moralities. But unlike Hobbes, Bergson also sees vital exigencies at work in critical rebellion against the limitations of such states of artifice.

The idea that there exists a dialectic of forces at work in nature is the fundamental argument of Bergson's earlier *Creative Evolution* (1907). Reflection on the world as revealed by modern natural science—that is, phenomena at different levels of complexity that exhibit diverse principles of organization and have emerged over time—gives rise to the hypothesis of a noumenal *élan vital*. What is at work in this creative evolution is not, however, some simple movement toward complexity, but what Bergson describes as a "law of dichotomy" and a "law of double frenzy."

In nature there is a constant splitting up, differentiation, or alienation, not just through the creation of competing individuals, but in the very processes of individuation or change; change itself is divided or polarized, each polarity exhibiting its own fit or outburst of development and operation. Phenomena are both nonbiological and biological, with the *bios* splitting further into plant and animal kingdoms, each of which operates more or less independently, while nonetheless feeding off each other. So too in human affairs there is a movement toward closed, material self-preservation—the movement of science-technology or what Bergson calls mechanism—and toward open relationship or unification—mysticism. Indeed, there is even a sense in which each of these movements in human affairs depends upon and is present in the other. Mechanism best flourishes not just as a means to material power but as a kind of mystical or ascetic commitment; for mysticism to flourish well there must also exist some level of mechanical satisfaction of material needs.

If our organs are natural instruments, our instruments are in themselves artificial organs. The tool of the worker is the continuation of the arm; the equipment of humanity is therefore a prolongation of the body.

Nature, in endowing us with an essentially fabricating intelligence, has therefore prepared for us a certain enlargement. But machines which run on oil, or coal . . . , and which convert into motion potential energies accumulated during millions of years, have actually given to our organism an extension so vast and a power so formidable, so disproportionate to its size and strength, that . . . what we need are new reserves of potential energy, this time moral energy. . . . The body now enlarged calls for a supplement of soul. . . .[16]

The *élan vital* divides itself, specializes, with each specialization developing on its own. But each must also continuously reach out to recover the other. The negative statement of the problem of the destructive relations of science and technology thus does not go deep enough. It is true that the industrial age has achieved a unique conquest of the material world, and in the process created special dangers for human life. But these dangers can be met by corresponding developments within morality, developments that are present and supported in hidden ways through science and technology themselves. "At the rate with which science advances, the day approaches in which adversaries . . . will have the means of annihilating each other,"[17] thereby creating strong incentives against warfare. Nuclear deterrence is anticipated by almost two decades. Moreover, "Whereas physics and chemistry help us to satisfy and thus encourage us to multiply our needs, one can foresee that physiology and medicine will reveal more and more clearly the dangers of this multiplication, and the disappointments in most of our satisfactions."[18] The health and fitness movement of the 1980s was forecast in the 1930s.

The phenomenal world is characterized by "oscillation and progress, progress through oscillation. And we should foresee, after the ever-increasing complexity of life, a return to simplicity."[19] The *élan vital* protects itself through and in the midst of its own diversification and change, and a return to the simple life, if not simply to life, is within the bounds of its noumenal cunning.

Independent of any noumenal grounding, however, it is possible to consider in more detail those positive phenomenal tensions that also can help overcome or break the connection between science, technology, and military. For this less speculative and more positive analysis of the science-technology-military relationship it is appropriate to turn to Husserl. The crisis of Husserl's *Crisis* is in the first instance "the crisis of the sciences" and their inability to answer "the questions of the meaning or meaninglessness of the whole of human existence."[20] Since the human existence in question is that of a European humanity informed by the positive sciences, this constitutes an "expression of the radical life-crisis of European humanity."

The historicophilosophical foundation of the crisis of positive science, as Husserl sees it, is that modern science originates with Galileo and the mathematization of nature. This mathematization involves a restricted thematization of human experience of the world, with the promise of positive answers not just to positive or empirical questions but also to metaphysical ones, or what Husserl calls questions of reason and meaning. The crisis is precisely that modern science is able to supply the former but not the latter, and yet its very successes appear to block access to any alternative approach to questions of meaning.

The positive character of the positive sciences, their fact and power orientation, and even their positive achievements, in conjunction with obstructed access to questions of meaning, can also be said to leave them open to militarization. A meaningless world of power relationships in which might makes right is precisely the world envisioned or thematized by the military attitude. The formation of modern natural science as an abstraction from the richness of "the natural, primordial attitude"[21] thus virtually entails militarization. It is therefore perhaps no accident that Galileo himself developed his new science in the service of more effective use of military artillery.

The irony for Husserl is that Galileo's reform of theory was originally an attempt to overcome the delimitation of experience that characterizes Western theory itself. Theory or philosophy originated among the Greeks when one aspect of human experience, simple wonder at the way things are, was taken out of "natural life" or "life naively, straightforwardly involved with the world"[22] and turned into a special theme, attitude, or project. Theory of philosophy itself takes one aspect of the primordial *Lebenswelt* or "life-world" and intensifies it, pursues it to the exclusion of the rest of experience, of the whole to which it is related and of which it is only a part. In the course of European history this project becomes in its development more and more alienated from the real world until it takes on the form of intellectual decadence in late medieval scholasticism. Galileo's attempt to redirect this pursuit back to the real world through mathematization is, in the first instance, an attempt to reconnect to the whole. But the whole with which reconnection is made is nevertheless already attenuated or deprived of its primordial richness precisely because of the means or method—mathematization. The resultant science and technology do indeed reconnect, but to a world deprived of richness and thus readily dominated, as if by the military. Yet within the *Lebenswelt* the abstraction or mathematization of nature remains only one possibility, and in reality there are dialectic alternatives open to recovery.

In the world in which we immediately live there is no science, no technology, no military. There are events or acts and experiences that might subsequently be thematized as belonging to the categories of knowing, of fabricating, or of fighting, but at the level of primordial experience neither individual nor community perform any such thematization. Hence in the primordial lifeworld there is, strictly speaking, no knowledge, no artifice, no war. Reflecting, indeed even incarnating this lifeworld with a certain naive clarity—in a presence that remains allusively alive through childhood memories, dreams, and the richness of our uninstitutionalized experiences—the archaic or preliterate cultures of the world also lack rigid distinctions between science, technology, and military. For the hunting and gathering peoples of the world, the initial differentiations of knowing, of fabricating, and of fighting remain amorphous, pregnant with transformations from one to another. At all times all members of the tribe are and are not, depending on ritual or seasonal context, seers, tool makers, warriors.

The law of noncontradiction has not yet entered into the lifeworld to manifest itself in firm differentiations. To be or not to be is not yet defined by institutionalized social roles and concomitant duties. The war paint of the warrior allows a person to take on not the institutionalized role of a warrior but what may be called the dramatic persona of a warrior, to become a warrior in a much more immediate but fluid or tentative sense. The child puts on the costume of a cowboy in the morning, of an Indian at noon, and of a doctor after dinner, without ever becoming cowboy, Indian, or physician.

But then the child also never ropes a red-blooded steer, kills a flea infested buffalo, or treats a life-threatening illness. Indeed, anthropologists have described the science, technology, and warfare of archaic peoples as like playing at being scientists, technicians, or warriors. Their science is a mixture of fact and fantasy, myth; they are hunters and gatherers, wanderers, who only pick at the ground and never become true farmers or the builders of cities; their wars are agonistic games, ritual engagements. According to Machiavelli's well-known remark, even late medieval wars "were initiated without fear, prosecuted without danger, and concluded without loss."[23]

Against this background the positive statement of the problem of the science-technology-military relation can be seen as an instance of a more general issue and perhaps even a foundation of the negative statement of the problem. Prior to institutionalized differentiation there are not any problems. Differentiation gives rise to problems, especially the problem of what can be termed the fallacy of reified differentiation, in which institutional roles take on an oppositional character and lose contact with, turn back upon, and undermine their origins. Lewis Mumford refers to the aboriginal form of such rigid, hierarchical, social organizations as "megamachines," and notes their regular affiliation with the military.[24] Indeed, through institutionalized opposition, the military role itself begins to pervade all other roles. The general form of a solution to this problem is to seek some kind of return to the source or recovery of nondifferentiation through articulated integration or by means of relations that encompass or transcend distinctions.

For example, the problem with the existence of knowing, of fabricating, and of fighting—rather than knowing-fabricating-fighting, which is neither knowing, nor fabricating, nor fighting—is that what is appropriate to knowing is different from what is appropriate to fabricating, and still again different from what is appropriate to fighting. The institutionalized differentiation of science, of technology, and of the military (or any other differentiation) introduces into human life, and the now historicized lifeworld, tensions between different ways of acting and the need to prosecute or defend alternatives. Such tensions must be overcome by being assigned their proper place in a larger whole. In Husserl's own words:

> In a restricted sense we call science, art, military service, etc., our "vocation," but as normal human beings we are constantly (in a broad sense) involved in many "vocations" . . . at the same time: we are at once fathers, citizens, etc. Each vocation has its time of actualizing activities. Accordingly, [any vocation] finds its place among other life-interests [and] has "its proper time" within the . . . form of the various exercised vocational times.[25]

The *Lebenswelt* is a variegated phenomenon that independent of any noumenal *élan vital* calls for both recognition and protection in its very diversity.

HISTORICAL SOLUTIONS TO THE PROBLEM OF SCIENCE, TECHNOLOGY, AND THE MILITARY

One trend toward a real-world resolution of the more negative form of the science-technology-military problem (the problem of scientific and technological world domination and destructiveness) has been indicated by historian and sociologist Derek de Solla Price. During the 1960s Price undertook to develop a metric analysis of scientific progress. One of the key characteristics of modern science as an institution is the scientific journal. The first scientific journal is the *Philosophical Transactions of the Royal Society,* which originated in 1665. Today, there are literally thousands of scientific journals. If the number of scientific journals is plotted against time, the resulting graph follows the general pattern of an exponential increase. Indeed, whether the number of scientists or cost of science is plotted against time, there results a sharp exponential curve.

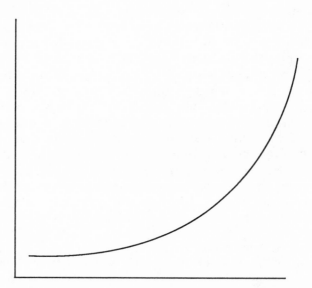

Since it is common to think of modern science and modern technology as closely related, it comes as no surprise that a metric analysis of the history of modern technology in terms of numbers of patents or numbers of engineers yields a similar curve.

What may come as something of a surprise, however, is that a metric analysis of warfare in terms of numbers of wars and numbers of deaths over a similar period—from 1700 to 1950—reveals the same basic curve. Moreover, in terms of amounts of money expended on the military worldwide, in a world that is undergoing global scientific and technological development, the curve even outstrips per capita Gross Domestic Product (GDP).[26]

Now the existence of similar curves in the metric analysis of the history of science, of technology, and of the military over the same historical periods again points toward a close and symbiotic relationship between science, technology, and the military. But it is possible to see further implications in these curves. To begin with, as is well known, the kind of exponential growth indicated by such a curve is a common characteristic of a host of inorganic and organic phenomena. For example, the rate of a chemical reaction often increases exponentially with rising temperature; and the number of both fruit flies and people in a freely reproducing population increases exponentially over time. But as is also well known, such exponential increases cannot go on forever. They approach saturation levels and undergo decisive transformations. Chemical reactions explode; population growth outstrips living space or available food, and social organization takes on a new form. The exponential curve gives way to a logistic curve:

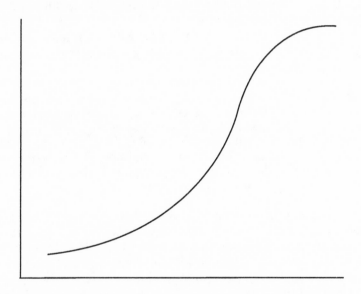

With regard to scientists and engineers, the numbers cannot keep doubling every fifteen years while population doubles every forty-five years, because there will come a point at which everyone is a scientist or engineer and the rate of increase will at least have to slow to that of the population. It is also the case that the amount of money spent on scientific and technological research and development cannot continue to increase faster than total governmental expenditures or GDPs.

Furthermore, as Nicholas Rescher has argued in an extension of

Price's research, it can be shown that major scientific discoveries are becoming increasingly expensive.[27] X-rays were discovered with a simple photographic plate, but the next subatomic particle may depend on the ten-billion-dollar superconducting super-collider. What Price calls Little Science, which underwent exponential growth, becomes mature or Big Science. Likewise with the military: its expansion simply cannot go on indefinitely at recent rates. Indeed, we can postulate that all three are approaching their limits or saturation points and that their growth is beginning to taper into a new institutional order.

This would constitute, especially for modern science, a unique situation. According to Karl Popper (and some other philosophers of science) modern science, in order to exist, must progress. Does this mean that if its progress slows down it is less modern or less science? Perhaps we are living in the presence of the origins of what could be called postmodern science and technology—and maybe even a postmodern military.

The central characteristic of the new form of science and technology, according to Price, is acknowledgement of the ideal of social responsibility.

> In the old days of Little Science there was tremendous reaction against political action by scientists. [Scientists] valued their independence; on the whole they liked *things* but were not very good at *people*. . . . [What is happening with the arrival of full logistic maturity in Big Science] is the maturing of a certain responsible attitude among scientists analogous to that which, in almost prehistoric times, moved physicians toward the concept of the Hippocratic oath.[28]

Price's observations from the mid-1960s have only been confirmed by subsequent developments. In the 1970s social responsibility, with its appeal away from science-technology to the human, became the primary theme in ethical discussions of science and technology,[29] and the most commonly invoked category for justifying or promoting delimitation of the influence of the military on science and technology or world domination and destruction by science and technology themselves.

There is a sense in which this idea of responsibility and the human is not only a reaction against what has been called the negative version of the science-technology-military relationship, but also grows out of the more positive analysis. The appeal to responsibility is commonly cast in the form of an appeal to transcend the institutional roles of a professional scientist or engineer. Institution in this context refers to any significant and persistent element in the life of a culture that is maintained and stabilized through social regulation, often by the state.

Science, technology, and the military have been transformed not just into significant and persistent elements in the life of our culture, but into dominant and mutually related elements funded and regulated by state bureaucracies, as well as by professional codes of conduct. But even more significantly, the pursuit of any of these institutions, and especially the pursuit of all of them together, can readily lead to the overwhelming of the human qua human.

The possibility of an opposition between the human and a social or institutional role is inherent in any differentiation of the human. Hence the argument for social responsibility commonly invokes the human over the role. "Do not act so much like a scientist. Act like a human being." "Do not always analyze things in technological terms, but consider the person." "Do not just be a soldier, be yourself." In its most general terms the appeal is simply to "be human." Any institutional role is reduced to what might be called a playful role, relativized, reflecting the postmodern recognition of the contingency of all definitions (or differentiations) of the human.

That such a criticism of science and technology as social institutions on the basis of an appeal to life has not been without real-world efficacy is confirmed by the thought and life of two paradigmatic scientists of the twentieth century, Albert Einstein (1879–1955) and Andrei Sakharov (1921–89). In an address to students at the California Institute of Technology in 1931, Einstein articulated an appeal to the human as the basis for his recurring ethical-political criticisms of science and technology especially in relation to the military.

> Why does . . . applied science [Einstein asks] bring us so little happiness? . . . In war it serves that we may poison and mutilate each other. In peace it has made our lives hurried and uncertain. . . . It is not enough that you should understand about applied science in order that your work may increase man's blessings. Concern for the man himself and his fate must always form the chief interest of all technical endeavors . . . in order that the creations of our mind shall be a blessing and not a curse to mankind. Never forget this in the midst of your diagrams and equations.[30]

Sakharov, too, as if taking up where Einstein leaves off, since the late 1950s made respect for human rights increasingly central to his criticism of the military and totalitarian abuse of science and technology. In his Nobel Peace Prize lecture of 1975 he speaks of the

indissoluble link between "peace, progress, human rights" and of
his conviction that respect for human rights "provides both the
basis for scientific progress and a guarantee against its misuse to
harm mankind."[31]

The exact character of the human in Einstein's and Sakharov's
conceptions is, however, what may be termed an empty conception,
one that calls not for development in accord with some substantive
understanding of human nature but simply for a nonviolation of the
human, whatever it might be. The human is conceived not so much
as a substance as a space. To restate the appeal with less vitality but
perhaps appropriate theoretical reform, one could venture the following: In a play there are what might be called the rights of the set
or the framework within which the play takes place. Actors must
respect, act responsibly toward, the stage on which they perform
their drama. Without such recognition of the rights of the set, a
drama will be unable to play itself out, will self-destruct. The
general stage on which scientific, technological, and military institutions can exist, it is argued, is that of human life. They must be
relativized or judged in terms of this life. Such is the ultimate
intimation of both Bergson (the *élan vital*) and Husserl
(*Lebenswelt*), and perhaps even of Einstein and Sakharov.

NOTE FOR A CRITIQUE OF THE APPEAL TO LIFE

The problem of the science-technology-military relation is that it
can destroy the stage on which it takes place. Recognition of this,
and of the historical tendencies to relativize commitments to science, technology, and military certainly deserves our support as a
helpful readjustment in human affairs. The implicit or ultimate
appeal to life as a kind of absolute good, however, calls for a more
cautious assessment.

Modern science and technology, and even the modern military,
were founded in the early modern period precisely on the basis of
an appeal to life. For Francis Bacon and René Descartes, modern
science and technology are preferable to traditional science and
technology because of their protecting and enhancing powers. Both
their medicinal and military utilities are specifically extolled. The
original philosophers of technology, from Ernst Kapp to José Ortega y Gasset, likewise present technology as what Oswald
Spengler calls "tactics for living."[32] From the romantic critique of
the Industrial Revolution through Bergson and Husserl to contemporary environmental ethics, the unifying theme in the critique of

modern technology is likewise its failure to live up to its life-protecting promises.

The original philosophic ideal, however, relativizes not just institutional roles of scientist, technologist, and warrior, but even life itself. For Socrates, it is not life and death but questions of justice and injustice or the good and the bad that provide fundamental guidelines for human existence.[33] Not physical life *(bios)* but a certain kind of life *(zoe philosophou)* is the final Socratic appeal. It remains to be determined whether the postmodern reaffirmation of life and delimitation of science, technology, and the military is ultimately philosophical or nihilistic in character.

When Hernán Cortéz landed at what is now Veracruz in 1519 and then advanced on to Tenochtitlan, the city of Moctezuma, he represented the leading edge of a military-technological assault on the New World. This assault culminated in the destructive obliteration of a three-thousand-year tradition of Meso-American civilization that had never developed the kind of heavy military emphasis characteristic of European culture. The Olmec, Zapotec, Teotihuacan, Maya, Toltec, and Mexican societies were based in religion centered cultures with highly developed craft and construction technologies—but virtually nothing in the way of military fortifications or weapons. Indeed, these civilizations even failed to develop such rudimentary technologies as the wheel (except in toys), perhaps precisely because they did not domesticate beasts of burden. Although they practiced human sacrifice, this aspect of a complex culture may well have caused less total intentional loss of life than the warfare of the so-called higher civilizations of Europe. Whether the unqualified intention to take control of human life away from nonhuman forces is capable of any unqualified promotion of life thus remains a fundamental question for the new world and its new technologies.

NOTES

1. See Carl Mitcham and Philip Siekevitz, eds., *Ethical Issues Associated with Scientific and Technological Research for the Military*, vol. 577 of *Annals of the New York Academy of Sciences* (29 Dec. 1989). This proceedings of a conference (26–28 January 1989) includes Mitcham's "The Spectrum of Ethical Issues Associated with the Military Support of Science and Technology" (1–9) along with thirty-one other contributions.

2. According to Quincy Wright in "The Study of War," vol. 16 of *International Encyclopedia of the Social Sciences* (New York: Macmillan, 1968), "The history of war can be conveniently divided into five great periods . . . distinguished by the technologies utilized in lethal conflicts" (455). This thesis is developed at greater

length in Wright's *A Study of War, With a Commentary on War Since 1942,* 2d ed. (Chicago: University of Chicago Press, 1965). According to William H. McNeill in *The Pursuit of Power: Technology, Armed Force, and Society since* A.D. *1000* (Chicago: University of Chicago Press, 1982), his "study of macroparasitism [complementing an earlier study of microparasitism in *Plagues and People* (Garden City, N.Y.: Doubleday, 1976)] among human populations [is] a study of the organization of armed force with special attention to changes in the kinds of equipment warriors used. Alterations in armaments resemble genetic mutations of microorganisms in the sense that they may, from time to time, open new geographic zones for exploitation, or break down older limits upon the exercise of force within the host society itself" (vii). Alex Roland's "Technology and War: A Bibliographic Essay," in *Military Enterprise and Technological Change: Perspectives on the American Experience,* ed. Merritt Roe Smith (Cambridge: MIT Press, 1985), 345–79, provides an excellent analytic overview of the literature. Five subsequent works of importance are as follows: Barry Buzan, *An Introduction to Strategic Studies: Military Technology and International Relations* (London: Macmillan and the International Institute for Strategic Studies, 1987); Everett Mendelsohn, Merritt Roe Smith, and Peter Weingart, eds., *Science, Technology and the Military,* 2 vols., *Sociology of the Sciences Yearbook 1988* (Boston: Kluwer, 1988); Geoffrey Parker, *The Military Revolution: Military Innovation and the Rise of the West, 1500–1800* (New York: Cambridge University Press, 1988); Robert L. O'Connell, *Of Arms and Men: A History of War, Weapons, and Aggression* (New York: Oxford University Press, 1989); and Martin van Creveld, *Technology and War, From 2000* B.C. to the Present (New York: Free Press, 1989).

3. For a brief review of various interpretations of this transition, see Melvin Kranzberg and Joseph Gies, *By the Sweat of Thy Brow: Work in the Western World* (New York: Putnam, 1975), 23–26.

4. Plato, *Republic* 2, 372cff. For comparison, see the preface to Niccolò Machiavelli's *Dell'arte della guerra:* "But in considering the ancient orders [we find that there is a close relationship between civilized and military ways of life] because all the arts, that have been ordered in a civilization for the common benefit of humanity, . . . would be in vain if no military force were prepared to defend them."

5, Thucydides, *History of the Peloponnesian War* 1, iii, 70–71.

6. See, e.g., Carolyn Merchant, *The Death of Nature: Women, Ecology, and the Scientific Revolution,* 2d ed. (New York: Harper & Row, 1989).

7. For relevant documentation and references on this point, see Daniel Greenberg, *The Politics of Pure Science* (New York: New American Library, 1967), especially book 2; and David Dickson, *The New Politics of Science,* 2d ed. (Chicago: University of Chicago Press, 1988), especially chaps. 1 and 3. Two key historical studies are: A. Hunter Dupree, *Science in the Federal Government: A History of Policies and Activities to 1940,* 2d ed. (Baltimore: Johns Hopkins University Press, 1986); and Jeffrey K. Stine, *A History of Science Policy in the United States, 1940–1985,* Science Policy Study Background Report no. 1, Task Force on Science Policy, Committee on Science and Technology, U.S. House of Representatives, 99th Cong., 2d sess. (Washington, D.C.: GPO, September 1986).

8. See, e.g., David Hume's essays "Of Commerce" and "Of Refinement in the Arts" (1741); Charles Montesquieu, *The Spirit of the Laws* (1748), especially vol. 1, book 20; and the articles on "Commerce" and "Luxury" in the *Encyclopedia or Rational Dictionary of Sciences, of Arts, and of Crafts* (1751–1765). William James's "The Moral Equivalent of War" is included in his *Memories and Studies*

(New York: Longmans, Green, 1911), 267–96. James, it should be noted, criticizes "trade" or commerce as lacking in sufficient discipline and nobility to be able to serve as a comprehensive substitute for war.

9. For Nobel (1833–96), see Erik Bergengren, *Alfred Nobel, The Man and His Work,* trans. Alan Blair (New York: Nelson, 1962): "Several times, especially after the invention of ballistite in 1887, he stated his belief that if explosives could only be perfected technically to a destructive and terrifying dimension, they would be a deterrent to war. The mere knowledge of their existence in such a form would, he contended, deter mankind from resorting to them as war weapons and thus promote peace. In this far-sighted thought, expressed as early as 1876, he was . . . far ahead of his time" (185–86). Jan Block (also Ivan Block and Jean de Bloch), *Budushchaia voina v teknicheskom, ekonomicheskom i politicheskom ot-nosheniiakh,* 6 vols. (1890), vol. 6, *The Future of War in Its Technical, Economic, and Political Relations,* trans. R. C. Long (New York: Doubleday, 1899), with many reprints and abbreviated versions. Timothy Garton Ash in "Ten Thoughts on the New Europe," *New York Review of Books* 37, no. 10 (14 June 1990), restates the issue:

> At the end of the 18th century Immanuel Kant suggested that the only states which would not necessarily go to war with each other . . . were those in which the "civic constitution" was "republican": that is, with limited government, the rule of law, and kings who listened to philosophers like him. . . . In 20th century terms liberal democracies don't fight liberal democracies. . . . [But] fortunately, that is not our only hope [for peace in Europe]. At the end of the 20th century there is another reason for states to avoid wars, one undreamed of in Kant's philosophy. This is that they have the power to wipe each other out. The development of military technology first allowed what Raymond Aron called the "eternal rivalry of states" to wreak untold carnage in "the century of infernal machines." But its further development then put a matchless check to that rivalry. (22)

The less pure view that technological weapons (and means of surveillance) will not so much eliminate war as simply make it controllable has been advanced by more establishment thinkers such as Henry Kissinger in *Nuclear Weapons and Foreign Policy* (New York: Harper, 1957).

10. For some relevant data, see, e.g., Ruth Leger Sivard, *World Military and Social Expenditures* (Washington, D.C.: World Priorities, 1974–present). The twelfth edition covers 1987–88.

11. According to William James (see n. 8) it is precisely because, when properly conceived, "the war against war is going to be no holiday excursion or camping party" (267) that such activity can call forth sufficient discipline and high ideals as to become "the moral equivalent of war." In light of "the aesthetical and ethical" evaluation of warfare (283), James does not believe

> that peace either ought to be or will be permanent on this globe, unless the states pacifically organized preserve some of the old elements of army-discipline. A permanently successful peace-economy cannot be a simple pleasure-economy. In the more or less socialistic future towards which mankind seems drifting we must still subject ourselves collectively to those severities which answer our real position upon this only partly hospitable globe. We must make new energies and hardihoods continue the manliness of which the military mind so faithfully clings. Martial virtues must be the enduring cement; intrepidity, contempt of softness, surrender of private interest, obedience to command, must still remain the rock upon which states are built—unless, indeed, we wish for dangerous reactions against commonwealths fit only for contempt. . . . The war-party is

assuredly right in affirming and reaffirming that the martial virtues . . . are absolute and permanent human goods." (287–88)

12. Henri Bergson, *Les deux sources de la morale et de la religion,* in *Oeuvres* (Paris: Presses Universitaires de France, 1959), 4, 8, 1210.

13. Ibid., 4, 13, 1217.

14. Ibid., 4, 13, 1216.

15. Ibid., 4, 13, 1217.

16. Ibid., 4, 28, 1238–39.

17. Ibid., 4, 14, 1219.

18. Ibid., 4, 24, 1231.

19. Ibid, 4, 22, 1230.

20. Edmund Husserl, *Die Krisis der Europäischen Wissenschaften und die transzendentale Phänomenologie,* in vol. 6 of *Husserliana,* ed. Walter Biemel (Haag: Martinus Nijhoff, 1954), sec. 2, 4, lines 10–11. English quotation adapted from *Crisis of European Sciences and Transcendental Phenomenology,* trans. David Carr (Evanston: Northwestern University Press, 1970), sec. 2, 6.

21. Husserl, *Crisis of European Sciences,* 281.

22. Ibid.

23. Niccolò Machiavelli, *Istorie Fiorentine* 5, i. For Machiavelli, of course, such "bloodless wars" are caused by the Christian emasculation of culture and are destined to be overcome by its own overcoming. See also *The Prince* 12 and *Discourses* 2, xvi–xviii. For a more affirmative appraisal of the same phenomenon, see Johan Huizinga, *Homo Ludens: A Study of the Play-Element in Culture,* trans. R. F. C. Hull (New York: Roy, 1950), chap. 5, "Play and War."

24. Lewis Mumford, *The Myth of the Machine,* vol. 1 of *Technics and Human Development* (New York: Harcourt Brace Jovanovich, 1967), especially chap. 9, "The Design of the Megamachine."

25. Husserl, *Crisis of European Sciences,* sec. 35, 136.

26. See Sivard, *World Military and Social Expenditures.*

27. Nicholas Rescher, *Scientific Progress: A Philosophical Essay of the Economics of Research in Natural Science* (Pittsburgh: University of Pittsburgh Press, 1978). See also Rescher's *Unpopular Essays on Technological Progress* (Pittsburgh: University of Pittsburgh Press, 1980), especially chap. 8, "Scientific Progress and the 'Limits to Growth'."

28. Derek John de Solla Price, *Little Science, Big Science . . . and Beyond* (New York: Columbia University Press, 1986), 100–101. This is an expanded edition of a volume first published in 1963.

29. See Carl Mitcham, "Responsibility and Technology: The Expanding Relationship," in *Technology and Responsibility, Philosophy and Technology* 3, ed. Paul T. Durbin (Boston: Kluwer, 1987), 3–39.

30. See the article "Einstein Sees Lack in Applying Science," *New York Times* (Tuesday, 17 February 1931). Brief quotation printed under the title "Science and Happiness," *New York Times* (Sunday, 22 February 1931).

31. Andrei D. Sakharov, *Alarm and Hope,* ed. Efrem Yankelevich and Alfred Friendly, Jr. (New York: Knopf, 1978), 5.

32. Ernst Kapp, *Grundlinien einer Philosophie der Technik: Zur entstechungsgeschichte der Cultur aus neuen Gesichtspunkten* (Braunschweig: Westermann, 1877); José Ortega y Gasset, *Meditación de la técnica* (from 1933), vol. 5 of *Obras Completas,* 3rd edition (Madrid: Revista de Occidente, 1943); and Oswald Spengler, *Der Mensch und die Technik: Beitrag zur einer Philosophie des Lebens* (Munich: Beck, 1931).

33. See Plato, *Apology of Socrates,* 28aff.

Scientific and Technical Development in a Democratic Society: The Roles of Government and the Media

MIGUEL A. QUINTANILLA

Science and technology policy is destined to occupy an increasingly essential and dominant position in today's democratic societies. Consequently, certain important changes will have to take place in the social institutions and in the characteristic habits of the representative democracies, in order to facilitate the participation of citizens in this field of politics and so maintain a legitimate capability for democratic institutions. This process should also be accompanied by important changes in public opinion, and this is where the media plays an important role. In analyzing science and technology policy within contemporary democratic society, I shall focus my discussion on the following points:

1. What are the specific problems posed by scientific and technological development in today's society?
2. What type of political responses can be given to these problems?
3. What role can be carried out by democratic institutions, especially government, in the design, assessment, and control of scientific and technological policy?
4. What is the responsibility of the media, specifically science journalism, in the field of science and technology?

SCIENTIFIC AND TECHNICAL DEVELOPMENT

Nearly everyone agrees that science and technology are decisive factors in explaining the dynamics of today's advanced societies.

Actually, since the industrial revolution, which began in the United Kingdom more than two centuries ago, we are accustomed to thinking that a society's progress depends on its industry, and that the latter advances through technical innovation, which increasingly depends on scientific investigation. Nevertheless, there are specific factors in the present situation that compel us to study in depth this now classic view of the connections between scientific-technical, economic, and social systems. These factors include: the extraordinarily rapid rhythm of technological change; the scope and depth of that change affecting all sectors of the economy and all social levels; and the close interdependence among technological innovation, scientific research, and social dynamics. Some important consequences, which I shall not explain in detail as they are well known to all, follow from these connections: scientific and technological development is one of the most important factors of economic growth; it is also a decisive generator in furthering deep, qualitative social changes that affect society as a whole; it is a process that greatly depends on human decisions. Indeed, we are now aware that we can enhance or restrict scientific research and technological innovation, and that what we do at this level will affect not only on our capacity for economic growth, but also the way that we grow and the resulting social transformations that we will experience.

POLITICAL RESPONSES

The perception of the universal nature of the challenge of technological change has caused many countries to adopt political response mechanisms to it. The assumption is that scientific and technical change is an essential variable in national economic development policies, and they have thus implemented new mechanisms and decision-making processes that permit public authorities to directly intervene in the promotion, control, and direction of technological change.

Political responses can be broken down into three types: promotion policies, orientation policies, and control and assessment policies. Promotion policies have as a main objective the creation of conditions conducive to the implementation of technological innovation (in any given country) with a rhythm and competitive intensity consonant with relevant international influences. The second type of policies includes those whose objective is not the indiscriminate promotion of scientific and technical development, but rather the orientation of development toward specific objectives defined according to the needs and resources of a given society, as well as in accordance with the relationship of technical development to the objectives of economic growth and qualitative development in that society. The third type consists of those political measures directed toward controlling the effects of technological innovation and forecasting possible lines of future development.

During recent decades, science and technology policy has expanded in the highly developed countries. At present, it seems reasonable to consider that an integrated policy of promotion, orientation, assessment, and control is the most adequate way of responding to the problems of scientific-technical development.[1] Clearly, there are notable differences among countries with respect to priorities and the emphasis that should be given to one or another type of policy. Nevertheless, I believe that any country seriously contemplating a scientific-technical development policy will have to confront the need to adopt an integral policy combining considerations of promotion, orientation, and assessment and control. For the purpose of formulating such an integrated policy, we can use a model of the flow of decisions affecting scientific-technical development in order to analyze the operations involved in an R and D program.

R AND D PROGRAMS

An R and D program is a plan of action the objective of which is to promote the investigation, the design, and the assessment of technologies. The basic supposition underlying an R and D program is that development of scientific and technological knowledge in a given area increases the possibilities of designing new technologies of interest to the organization that sponsors the program. Currently, most scientific research, and major technological innovations, are conducted through R and D programs. As a result, most political decisions in this area of science and technology evolve around the definition, implementation, and assessment of this type of program.[2]

An R and D program always responds to social objectives, as depicted in fig. 1. These, in turn, are determined, on the one hand, by the needs and wants or aims of the promoting group, and, on the other, by the scientific and technological resources available. The social and scientific-technological contexts of the program range from research carried out within a private company or group of

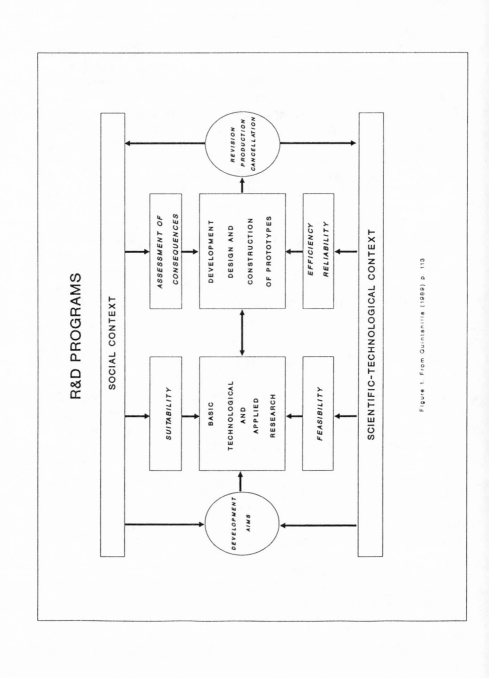

Figure 1. From Quintanilla (1989) p. 113

companies, to research carried out within the context of national scientific and technical development policies, to research conducted on an international scale. The determination of an objective conditions the elaboration of an R and D program, which involves three types of activities: research (basic and applied), development (system design, construction of prototypes), and assessment.

There are two types of assessment criteria involved in an R and D program, which we might call internal assessment and external assessment. Internal assessment refers basically to the scientific and technological viability or feasibility of the program, and to its intrinsic technological value as determined by such criteria as efficiency, effectiveness, and reliability. This so-called internal assessment is thus primarily of a scientific and technical nature, although its results may vary according to a given industrial or commercial point of view. External assessment of a technology may be of two kinds, depending on whether it refers to the properties of the technology, or to the consequences deriving from its use or application. In the first case, we can speak of the suitability of a technology or a technological application; in the second, of the impact or the consequences of this application.

An assessment for suitability can be carried out on an existing technological development already proven feasible and efficient, or on the projected objectives and the partial results of an R and D program. In the first case, this would really be a utility assessment that can be carried out through a cost-benefit analysis. In the second case, the utility forecast for a given technological development may become altered, once the feasibility research has progressed, as the efficiency, effectiveness, reliability, and safety values have been determined. Thus, at this level, the external assessment of a program closely depends on the internal assessment and is, in practice, a process subject to continuous revision.

Assessment of the consequences refers to the specific uses of a technology. With an available technology this means assessing the possible consequences of its application by a specific group under certain given circumstances. Thus, it is technological projects themselves that are, in this case, subjected to environmental impact controls and risk analyses. In a good R and D program, the assessment of consequences is carried out right from the design phase and is aimed at judging the consequences of the potential applications of the system in a wide range of possible circumstances. Three main types of criteria for assessing consequences can be defined: risks, environmental impact, and social impact. The risk associated with the application of a technology refers to its possible pernicious consequences for human lives, health, or the welfare of the potentially affected population. The risk assessment consists of calculating the product of the probability of occurrence of undesirable consequences, and the utility value (disvalue or cost) of these consequences. Thus, a subjective factor intervenes in risk assessment: the evaluation of potential liability to human life and health, for example, which at times makes it difficult to establish a rational assessment procedure.

Environmental impact assessment refers to an evaluation of the consequences that the application of a technology could have on the physical environment in which it is implemented. Impact can refer to any of the relevant variables that define an environment, from physical variables (geological, chemical, biological, atmospheric) to aesthetic variables (impact on the landscape). The ecological perspective is the one most frequently adopted for environmental impact analyses. This primarily entails establishing the point at which the introduction of a new technology in a specific habitat irreversibly alters the conditions of ecological equilibrium. However, environmental impact assessment is not limited solely to the immediate physical environment.

The assessment of social consequences is destined to have increasingly greater relevance, owing to the importance of new technologies in all sectors of society. Information and communication technologies, and their effect on employment, leisure, culture, and industrial organization are prime examples, but many other technologies can have considerable social consequences. Think, for example, of the consequences of the introduction of the railroad in the nineteenth century, or of the automobile in the twentieth century, or of the construction of large dams in rural areas. The specific problems posed in the assessment of social consequences (in addition to those that are common to risk and environmental impact assessments) are derived from the breadth and the lack of definition, of the possibilities that need to be considered, and from the absence of a stable point of reference. The truth is that any technology of some relevance will alter, to a greater or lesser degree, social structure, customs, and daily life. Unfortunately, in contrast to risk or environmental impact assessments, where it is assumed that reference values can be objective (the health or welfare of the population potentially involved, ecological equilibrium), nothing similar exists in the case of the assessment of social consequences. Even with the knowledge that the introduction of a new technology will have decisive effects on a social structure, the evaluation of these effects cannot be made by reference to a previously established objective criterion, unless it is assumed, in principle, that any social change is undesirable, in which case the only valid conclusion is that any technological change would also be undesirable.

The inherent difficulty in assessing technological consequences does not diminish the importance of assessment in technological development, but it does suggest that we must revise simplistic approaches to the task at hand. For example, it surely seems unreasonable to wait for problems to be resolved by means of simple calculation techniques. At the same time, there is a growing conviction that attention should be focused on processes in which the whole of society can participate in technological assessment. The adoption of such approaches seems to be increasingly prevalent. Thus, external assessment of technology increasingly acquires an unavoidably political dimension.

TECHNOLOGY ASSESSMENT AND POLITICAL DECISIONS

The expression "technology assessment" originated with the initiative to create the Office of Technology Assessment (OTA) in the U.S. Congress. The initial objective was to create a service that could inform Congress of the consequences that were likely to result from the introduction or the development of new technologies. Although the initial achievements of the OTA did not measure up to the hopes with which OTA had been invested, the initiative did contribute to a more precise definition of the methodological, political, and institutional problems associated with technology assessment. Presently, the experience of the OTA itself and other similar institutions created in many countries, at various governmental levels as well as on a private basis, has given rise to a better understanding of, and an extensive literature on, the assessment problems that must be confronted and the methods by which to do so.[3]

A broad consensus has been reached on the work involved in technology assessment understood as an academic specialization. The basic idea underlying these studies is that "it will be easier to direct technological development if investigations on the effects that a technology can have on society are carried out from the moment it is introduced."[4] Starting with this consensus, the perception of technology assessment has evolved and different definitions have been given. Currently, two main conceptualizations can be distinguished: reactive and active (or constructive).

In the 1970s, a reactive approach predominated in technological assessment, functioning as an "early warning system," the objective of which was to forecast possible undesirable social consequences related to the introduction of a new technology in comparison with existing alternatives, so that the agents responsible for making decisions could have maximum information in order to take corrective measures. Today, there has been a conceptual change toward a more active attitude, in which technology assessment focuses more on social problems, and the possible responses that technological development can provide, than on merely attending to the disturbing consequences of existing technological developments.

The current institutionalization of technology assessment shows the complexity of social interactions that influence technological development. In fact, technology assessment institutions are hybrid organisms. They have a scientific nature, but they serve political interests, and the working methodology they have developed (of a multidisciplinary, open, prospective, and participative nature) is a good example of the way to confront the new challenges that technological development poses to political activity with new methods and strategies.

I believe it is precisely this sort of social assessment of scientific and technological options that should be utilized in efforts to increase civic participation in the control of scientific and technical development. There are two basic mechanisms in democratic systems to increase civic participation in politics: legislative institutions and public opinion. It is important to understand the consequences that derive from involving legislative activity and the media.

THE ROLE OF THE LEGISLATURE IN SCIENTIFIC AND TECHNOLOGICAL POLICY

In Western democracies, the legislature or Parliament is formally the place where the political will of the people is expressed through complex representation mechanisms. The critics of representative democracies certainly have many reasons to complain: in the first place, the mechanisms do not always function adequately; in the second, even if they did, the representation of political interests and options owing to their very nature could never be the same as direct participation. Nevertheless, it seems that the best way of democratically organizing collective life is to base it on the typical mechanisms of representative democracies. Democracies based on legislative representation are precisely those that best guarantee the possibility of trying out new ways of participation and legitimizing those that obtain the best results.[5]

There are, however, other types of criticism and analysis of the role of legislatures that are relevant to the subject being discussed. In fact, many political theoreticians have pointed out the change of role that occurs in legislatures as a consequence of the change that has occurred in the nature of political problems themselves, and in the mechanisms of state administration. Simply put, and with a certain exaggeration that helps to put into relief the essence of the matter, it can be said that while today's politics have an increasingly technical and complex nature, parliamentary mechanisms continue to reflect their nineteenth-century models. In the best of cases, they are assemblies in which the elected representatives, using common sense and ideological premises, make decisions on political questions related to highly complex matters whose dominion is not usually within the scope of their professional competence.

One of the consequences arising from this situation is that the traditional mediation of social interests through legislative representation is inevitably superseded by the mediation of the political parties and, even more so, by the predominance of the information and technical competence of the administration at the service of the executive power, or at the service of the experts who work for powerful lobby groups. Surely, this is one of the basic causes of the displacement of the center of politics from the legislative arena to the executive branch, as well as of periodic suspicions that arise with respect to the capacity of legislatures to carry out efficiently their role in modern complex societies.

Personally, I do not agree with this type of analysis of the legislative role in today's politics. Moreover, I do not think that the basic problem is the displacement of the center of interest from the legislative to the executive branch, or from the individual members of parliament to political groups. I would stress that there is no real reason to think that this change in legislative activity is a consequence of, or a response to, the complexity of contemporary political problems. It is true that the executive branch, or lobby groups, may have greater information available to them than the public's elected representatives and greater technical expertise; but it is also true that not even the most copious information can substitute for political will when making decisions under conditions of uncertainty and risk.

In reality, the political decision-making processes have their own logic in which information and technical assessment is a necessary but never a sufficient condition. The basic problem raised in the analysis of political decisions in highly complex technical matters is not that of substituting political decisions for technical directives, but rather of guaranteeing that the information available to make the political decision is pertinent to the problem at hand. When this does not occur, the result is that decisions are adopted with inadequate criteria, independent of whether the responsibility lies with individuals or with the group.

The problem, then, is not so much the unsuitability of formal legislative mechanisms, given the technical character of present political decisions, but rather the fit of the instruments and the criteria that intervene in the decision-making processes to the nature of the matters to which these processes refer. This is especially relevant in the sphere of science and technology policy, where the most important decisions are strictly political and not technical, although the content and information required for making these decisions are eminently technical. Nevertheless, the specific and novel nature of the pertinent information is far more essential than its technical complexity.

My hypothesis is that, contrary to what occurs in other spheres of legislative competence, the legislative is the ideal place to make decisions regarding the social evaluation of technological development. To do this, however, legislatures need a specific type of information, which, as institutions, they should be able to generate. Without going into a description of the characteristics this information should have and how legislatures might organize to acquire it, I shall only point out some requirements that seem essential.

1. In the first place, legislators should have objective information readily available on the most important scientific and technological options.

2. In the second place, the information should attempt to be comprehensive, its collection guided by criteria based on feasibility and effectiveness, as well as by external assessment criteria based on suitability and social consequences.

3. Lastly, the process of elaboration and dissemination of the information collected should be participative and should specifically allow legislators to follow the process in conjunction with experts, and with those social sectors most likely to be affected by the political decisions to be adopted based on the information obtained.

As it currently stands, in spite of existing organizational and institutional differences, the various legislative experiments currently being carried out do seem to be oriented toward this direction. I believe that the spread of these types of legislative initiatives and instruments is going to continue, and in the future an integrated approach to scientific and technological policy will expand, just as promotion and orientation policies expanded in the 1960s and 1970s through the creation of specialized departments within the executive branch.[6] However, even with legislative institutions serving as the basic mechanism for political participation in representative democracies, there is as important role for the media as an instrument for the formulation and dissemination of public opinion on science and technology issues.

THE RESPONSIBILITIES OF SCIENCE JOURNALISM

In a pluralistic and democratic society, the existence of an adequately informed public with the capacity to express itself and freely exchange information is as important as the existence of democratic political institutions. This axiom of liberal democratic tradition is so obvious that it seems superfluous to repeat it. Nevertheless, it is not out of place here, in an attempt to see how the media and public opinion can influence this new field, how, in order to be effective, an integrated approach to science and technology policy requires active collaboration with, and a certain change of style and attitude on the part of, traditional science journalism.

If the importance of science journalism were measured by the space it occupies in a daily newspaper or on television, or in the number of weekly magazines published in most countries, one would have to conclude that this is a secondary area of interest for public opinion. Nevertheless, it must be pointed out that important differences do exist between some countries, that the importance of science journalism is proportional to the level of economic and cultural development of a given country, and that, as has occurred with economic information, scientific and technical information seem to be acquiring more and more relevancy.[7] However, the predominating styles and attitudes of current science journalism are not the most appropriate to promoting public participation in the assessment and control of scientific and technical development. At the risk of simplifying, and of being unjust with respect to other very positive values that undoubtedly exist in scientific journalism, it is important to recognize that the media uses some very common misconceptions and stereotypes when dealing with scientific and technological subjects, which unfortunately devalues their potential social and political role. In particular, I am speaking of scientific mystery, technological determinism, and journalistic social and political naiveté.[8]

Currently, scientific theories are generally presented almost as miraculous discoveries, incomprehensible to the layman, but of supposed great importance for humanity, although often without any stated reason. What stands out in such reporting is the sensationalism of the discoveries instead of the difficulty of the background work involved, and the importance of the problems that these discoveries may help to resolve. Thus, it is not infrequently the case that the reporting of scientific information is plagued with errors, imprecision, and exaggerations,[9] apparently on the grounds that the complexities of the problem do not interest the reader, who is assumed to be unable to understand scientific facts.[10]

Another dangerous misconception is that of technological determinism. This is apparent above all in the way that scientific and technical advances are presented as necessary products of our times, instead of as the result of a consciously assumed enterprise that consists of looking for particular technical solutions to particular social and human problems. Thus, the reading public is assumed to have a passive or merely contemplative attitude toward problems of technological development, instead of being capable of taking an active participatory role in solving those problems.

Lastly, the media often reflect a certain naiveté when dealing with social and political dimensions of scientific-technical development. Here, I refer, not to the necessary limitations imposed by the very nature of some forms of the media, but to two inaccurate but dominant versions of the reasons for and the values of science and technology: the optimistic version, in which all discovery is positive, and the pessimistic version, in its various forms, in which technological development leads to self-destruction, in which science serves capitalism, and any scientific-technical progress is a backward step as far as the emancipation of humanity is concerned. What the media misses by adopting either of these two positions is the important understanding that the level of scientific research, and the harm or benefits of technological development, depend on human decisions over which the public can have a determining influence.

I believe that in order to have conscious and conscientious participation by citizens, in the control and assessment of science and technology policy, the media must attempt to overcome these stereotypes and limitations. It will thus be essential to do the following:

1. Focus on scientific research and technological development as processes, instead of only on the results of these processes (scientific discoveries and new technical artifacts).

2. Present scientific-technical development as a social phenomenon that depends on human decisions and has an economic cost, social consequences, and a political dimension.

3. Stress the open nature of important questions with respect to the objectives of technological development, costs, the assessment of consequences, and promote a journalism of opinion on these matters, thus contributing to the creation of a democratic consensus.

4. Connect the information with national and international policies on scientific and technological development favoring the participation of citizens in debating this policy.

SUMMARY AND CONCLUSIONS

Science and technology are going to occupy an increasingly dominant place in modern societies. For this reason, science and technology policies will be more and more important, and the roles of assessment and democratic control of technological development will have an increasingly decisive role in social change. Consequently, democratic systems should be adapted to guarantee the adequate functioning of traditional mechanisms of representation and political participation in this area. Specifically, legislatures should have appropriate instruments to obtain objective, relevant, and comprehensive information on the problems of scientific and technological development, which would allow politicians to make decisions from an integrated perspective that draws upon rational and participative processes. The media has an important role to play in this enterprise, but science journalism will have to make an effort to adapt to the new demands by overcoming old stereotypes. It should encourage debate on the important options in national science and technology policy issues, on the consequences of technological development, and on the creation of a democratic consensus for making science and technology policy choices.

NOTES

1. The recent report of the Organization for Economic Cooperation and Development (OECD), *New Technologies in the 1990s* (1988), which considers technological innovation as a social process, makes recommendations typical of an integral policy.

2. Miguel A. Quintanilla, *Tecnología: Un enfoque filosófico* (Madrid: FUNDESCO, 1989), chap. 6.

3. The most complete and updated information is found in the Acts of Congress of Amsterdam on Technology Assessment: S. C. Hoo et al., eds., *Technology Assessment: An Opportunity for Europe* (The Hague: Dutch Ministry of Education and Science, 1987). In Spain, the journal *Telos* 12 (1987–88) published a very useful and updated dossier on technology assessment, "La evaluacíon de tecnologías," ed. A. Castilla et al., with papers from A. Castilla, M. Ross, L. Sanz, H. Tuininga, and M. Procter. In M. A. Quintanilla, ed., *Evaluación parlamentaria de opciónes científicas y tecnológicas, Seminario internacional* (Madrid: Centro de Estudios Constitucionales, 1989), speeches from the International Seminary on Parliamentary Assessment of Scientific and Technological Options held in the Centro de Estudios Constitucionales (Madrid, 20–21 April 1989) are included.

4. R. E. H. M. Smits et al., *The Possibilities and Limitations of Technology Assessment* (The Hague: Dutch Ministry of Education and Science, 1987), 2.

5. This evaluation of representative democracy will appear insufficient to some political philosophers. My own view on the question is developed in M. A. Quintanilla and R. Vargas-Machuca, *Utopía racional* (Madrid: Espasa Calpe, 1989).

6. The last recommendation of the 1988 OECD report (see n. 1) endorses this forecast. The authors state: "We therefore recommend the further development of various forms of technology assessment which should be a continuing process and in which elected legislatures should equip themselves to play an active and informed, though not an exclusive, role. The basic aim should be to provide information to those concerned, to promote and participate in a constructive public debate in a wide circle of institutions, thereby strengthening the democratic process through increased public understanding of, and involvement in, the process of change" (78).

7. A very interesting evolution has taken place in this respect during the last few years in Spain. At the end of the 1970s there was no specialized press for scientific information, and the general information media hardly conceded space to scientific-technical matters. Today, there exist several monthly scientific and technical magazines, which are orientated toward the general public and have ample diffusion. Recently, a weekly scientific and technological magazine appeared in Madrid, and all daily national newspapers have been publishing, for years now and with more or less success, weekly supplements of scientific and technological themes, generally accompanied by health and environmental themes.

8. Although I do not believe that these stereotypes are exclusive to the Hispanic cultural environment, there is no doubt that it has deep roots in Spanish folklore: all of them are expressed in this lamentable verse of a famous Spanish zarzuela, *La verbena de la paloma,* where Hilarion, the pharmacist, boasts of his knowledge by stupidly repeating, "Today . . . science is advancing in a way that's barbarous, brutal, bestial." Contrary to the national and racial genius of Hilarion,

the British Pygmalion used his scientific knowledge to make the girl from the slums literate.

9. The recent example of cold fusion is a good illustration; it seems that sensationalist journalism has contaminated the scientists themselves.

10. In this context, it should not seem surprising that the line between science and pseudoscience is increasingly more ambiguous: UFOs sell more magazines than communication satellites.

Science and the People: Science Museums and Their Context

ANTONIO TEN

THE MODERN CITIZEN AND TECHNOLOGICAL CIVILIZATION

The Western world is currently immersed in a technological ocean, the extent of which it is aware only up to a certain point. Citizens, even of the First World, living in a consumer society and impressed only by the greatest of discoveries that get coverage in the mass media, are largely unaware of the technified world to which they belong, and in which even their most insignificant activities take place. For the majority of these citizens, even for those who can be considered "cultured," television is a kind of magic window that they know how to turn on and off or change at will; the kitchen tap is a source of chlorinated water that comes out hot or cold, as you wish; the airplane is a mysterious machine that flies (better not to worry about exactly how it works). All problems involving energy consumption, transportation, health, nutrition, and leisure are resolved by means of mysterious instruments that somehow, in efficient and amazing ways, to which access is limited only by personal financial resources, satisfy the "dire necessities" of life.

The increased technological complexity of goods and services results in an unavoidable specialization, with the results that ever-smaller numbers of people are responsible for technological innovation, management, and control. The scientific and technological substratum on which our current civilization rests has become, for most, an esoteric world in which the technician, who resolves all problems, acts as the high priest, scientists and engineers who develop the most essential devices have become almost science fiction-type figures, and, in some cases, in the area of medicine for example, they have become almost demigods who are expected to produce miracles "a la carte."

Thus, the more scientifically and technically advanced society has become, the more the scientific-technological paradox, as it may be called, manifests itself. That is, the more science and technology the average citizen uses in daily life, the more incapable he or she feels of understanding the world in which she or he lives. This is a situation with no direct parallel in past eras. Scientific and technological culture and values have been superimposed on traditional culture that is still enforced within the family unit and throughout the educational system. The scope and potential danger of this superimposed culture is largely unrecognized by the majority of people. The tragic gap that exists between these two cultures, a distance that increases with each new advance in technology and after each new scientific discovery, separates the average citizen more and more from any possible understanding of his or her own world and from the possibility of controlling the consequences of his or her activities. Pollution, exploitation of natural resources, waste, and personal and collective stress are immediate consequences of these cultural and technological imbalances inherent in the current advanced postindustrial civilization.

Nevertheless, these same scientific and technological achievements have allowed our society to reach the level of material well-being that we presently enjoy, a level inconceivable only a hundred years ago, and one that we would apparently rather increase than abstain from. Except for a certain marginal sector, progress has become synonymous with the encouragement of a technological civilization, of the consumption of goods and services of an increasingly more elaborate nature.

Is it thus inevitable that, except in the case of those special citizens who have been initiated in some of the mysteries of our civilization, common mortals must assume the role of ignorant and acritical consumers of goods? Is it necessary that entire countries, and not only Third World ones, must be subjected to this state, or relegated to the role of mere suppliers of raw materials and non-specialized services? Although the logic of the world economic system would seem to oblige us to answer both questions affirmatively, universal rights to culture, information, and freedom, and the right to criticize, are values that contrast drastically with the deepening of the gap between "initiated" and "noninitiated" and would suggest that alternative courses of action are possible. It is thus important to look for ways of overcoming the ever-increasing differences between popular culture and the technological civilization, and to make available to the average citizen all the information required in order to understand the civilization in which he lives.

This is without doubt the biggest challenge that those who are responsible for citizen education in the year 2000 will face.

The middle-class family, unable or unwilling to take on this educational challenge, because of the generational gap and the rapidity of innovation, has resorted, as in the past to a classical solution: the school. The general public thus must acquire in the schools the knowledge they need to understand the world of which they form a part, in order to be able to get along successfully in it, and to participate actively in shaping the course of their own development. Unfortunately the school system, be it at elementary or secondary levels, has been experiencing, since the beginning of this century, an increasingly marked contradiction between the need for specialization in all areas—especially with respect to science, which begins earlier and earlier—and the classical ideal of a general liberal education.

The educational systems in today's Western society have revealed certain almost inevitable limitations. Having outlived a time of rote memorization of mostly literary and theoretical content, there is a need for a more practical formation that, beginning at an earlier point in time and backed by new educational techniques, requires a constant supply of new cultural and educational instruments. The size, complexity, and rapid evolution of the problem results in an inertia that often ends up preventing any new resources that are found from ever actually reaching the centers of need.

A second, not totally unrelated reason behind the need for alternative institutions for science education is that most citizens abandon studies in science fairly early, either in favor of other types of studies or in order to join the working world. Without an institutional mechanism adaptable to their level of interest, this large sector of the population soon loses the possibility of following the course of scientific and technological advancement, and with it, the capability of understanding, even at a general level, the mechanisms present in their world and the ability to judge their utility or suitability.

The school system, even though it has undergone transformations in order to adapt itself to the necessities of technological society, is not by itself enough to overcome the scientific-technological paradox. As long as the education system thus fails to provide a sufficiently flexible means of adjusting to the requirements of an evolving technological society, it is obvious that complementary specialized institutions are needed to take these aspects into account and provide for education in a more adequate manner.

SCIENCE MUSEUMS

Many societies, aware of the problem just posed, have, from the beginning of the twentieth century, made ambitious attempts to solve this need for new educational approaches for understanding scientific and technological culture by creating so-called science museums.[1] These museums of science, science centers or science and technology cities, as the new generation of these institutions are starting to be called, are informative and educational institutions of an essentially informal, noncompetitive, and nonregulated character. At these centers, the visitor, more curious than studious, should be able to find clear answers to questions he or she has formulated, consciously or unconsciously, on the subject of the surrounding technological civilization. The visitor also should receive enough stimulation to the imagination that he or she is able to envision the future as it might be for him or her and for what it may hold in store for his or her children.

From a historical point of view, the idea of science and technology museums goes far back in time. From the moment it became evident that mechanization was to play a fundamental role in the development of European society, and the necessity to communicate technical knowledge in order to improve industry and craft was felt, initiatives to create science and technology museums were undertaken.

Although a number of early precedents can be found among Renaissance scientific societies—for example, the Academia del Cimento, sponsored by Leopoldo and Ferdinand II de Medicis, which operated between 1657 and 1667, and several others whose masterpieces can be admired in the Museo de Storia della Scienza in Florence—the socioeconomic atmosphere resulting in the science-technology museum initiatives being discussed here is very different from these. The mentality that allowed for the appearance of institutions comparable to the scientific museums we have today can be traced to the Industrial Revolution in England during the change from the seventeenth to the eighteenth century. An undeniable cornerstone in the process of social awareness of the importance of technology and science in western Europe was the publication, in France, between 1751 and 1772, of the *Encyclopédie, ou Dictionnaire raisonné des Sciences, Arts et Métiers*. This key work by Diderot and D'Alambert, a truly universal monument to science and art, contributed decisively to the transformation of "mechanical" or "servile" arts into one of the essential components of culture during the Enlightenment. The honor attributed by the scholars to the "useful sciences" allowed industrial and artisan techniques, until then considered degrading by society, to become the object of study and protection by nobles and even the king.

The Enlightenment produced the first models that can be considered direct precedents of our science and technology museums. The first important public initiative was taken by Jacques de Vaucanson, who, in 1741, was named chief inspector of silk manufacturing in France. Vaucanson was constantly preoccupied with the idea of improving industrial techniques, and with that in mind, he assembled a series of the most interesting machines of the times, along with his own inventions, in the Hotel de Mortagne in the Rue de Charonne in Paris. Upon the death of Vaucanson in 1782, the collection was bequeathed to Louis XVI and, following a report by Joly de Fleury, Alexandre Vandermonde, a member of the Academie des Sciences, was noted as the first director of the Royal Machine Deposit. This collection was incorporated into the Conservatoire des Arts et Metiers at its creation on 10 October 1794. It has continued, along with the Musée des Techniques, to be conserved and extended up until our own time.[2]

The scholars of other countries understood very well the utility of these types of installations for their projects for popularizing new arts. Almost coincidental with the appearance of Vandermonde's Deposit, the Spanish Real Gabinete de Maquinas was created in Paris on the initiative of Agustín de Betancourt, who had received a commission from the first secretary of state, José Moñino, for the formation of a board of civil engineers after the fashion of the French École des Ponts et Chausées. At Betancourt's request, drafts and models of the most interesting machines in the École and those that were known elsewhere in France, the Netherlands, and England were copied and fabricated. The Spanish ambassador Count Fernan Nuñez visited the collection in 1788 and proposed the creation of one in Madrid to King Charles III. Betancourt worked on this project in Paris between 1788 and 1791, at which time developments in France seemed to advise the urgent transfer of the collection to Madrid. On 1 April 1792, in the Palace of the Buen Retiro, the Real Gabinete de Maquinas was opened to the public, with Betancourt in the position of director and, in his absence, Juan de Peñalver as assistant director. Peñalver was responsible for the Catalogo of the collection, published in Madrid in 1794. Unfortunately, Spain's tragic history in the years that followed resulted in the disappearance, without a trace, of the ingenious artifacts that Betancourt and Peñalver had assembled.[3]

From the time of these first initiatives up to those of the scientific-technical museums of our century, new experiences and, above all, new outlooks emerged. After the French Revolution and throughout the nineteenth century, the foundations for what would later be called the second industrial revolution or the scientific-technical revolution of the late nineteenth and early twentieth centuries were laid. News of the new technologies that were springing up in Europe and America quickly spread. At the same time, the emerging universal right to an education gave the newly educated classes access to scientific and technological information. These two elements, and the initiative that was already present at the beginning of the twentieth century, gave rise to the most important scientific-technical museums that exist today. To cite only the several most outstanding, in 1903 the Deutsches Museum was founded in Munich as a cultural and educational institution,[4] soon to be followed by the Science Museum, heir to the nineteenth-century South Kensington Museum of London, and by the Permanent Exhibition of Agriculture and Industry in the Soviet Union in Moscow. Numerous other science museums, including the North American Science Centers, were added to these, especially after World War II.

Through the experience accumulated on both sides of the Atlantic, new forms of teaching capable of extending science to the general public subsequently emerged.[5] In the past few years, there has been a veritable boom with regard to this type of establishment—for example, to name only a few recent ones, the Cité des Sciences et l'Industrie in Paris, opened in 1986; the Museum für Arbeit und Technik in Mannheim, Germany; the Heureka, the Finnish science center in Vantaa, near Helsinki, Finland; and the Exploratory Science Center in Derry, Great Britain.

TWO PHILOSOPHIES

Even though a wide-spread consensus has been reached regarding interest in and the need for such scientific and technological educational institutions, no standard model has been formulated, and existing centers currently face a dilemma posed by two completely different philosophies of purpose.

The first philosophy, of which I have cited two examples, might be called the "scientific museum." These museums may include scholarly collections of irreplaceable pieces of scientific craftsmanship, as well as a strong interest in technology, in addition to their fundamental interest in science, which leads to the desire that the visitors to the museum encounter specific pedagogical experiences and active experimentation. Such is the concept of the Musée des Techniques in Paris, the Deutsches Museum in Munich, the Science Museum in London, the science and technology museums administrated by the Smithsonian Institution in Washington, and many other museums, especially in the area of technology, all over the world.

The second, that of the "science center," as they are known worldwide, stems from a completely different philosophy. In 1937, Paris celebrated the Exposition internationale des arts et des techniques, an international exhibition that covered twenty-three thousand square meters of the Grand Palais, at the center of which was the Palais de la découverte. The Palais de la découverte opened up an immense area of scientific experience to the general public, in which the laws of nature were directly represented as experiences susceptible to manipulation, explained by specialists for the benefit of the visitors. A similar idea was formulated and exploited at the end of the 1960s in the United States and Canada, and from there the notion of science centers evolved quite rapidly and broadly. One of the most widely known models is the Exploratorium of San Francisco, created by F. Oppenheimer at the end of the 1960s[6] and which was copied dozens of times across the United States and in the Ontario Science Center in Toronto, Canada. The philosophy behind these centers was rapidly extended to the rest of the world, especially to those countries under Anglo-American influence, so that there are now splendid examples of these types of installations in Singapore (the Singapore Science Center), in Japan (the Yokahama Science Center), in Australia, in Hong Kong, or in centers such as Heureka in Vantaa, inaugurated in May 1989. In Europe, England was the country most involved in this movement and demonstrated its conviction by opening the Exploratory in Bristol, the Techniquest in Cardiff, and the Heureka in Halifax.

France, with its splendid Cité des Sciences et des Techniques in Paris, better known as La Villette, has contributed to this movement with an original work in which the building, architecturally innovative and spectacular, combines a science center with an industrial exhibition. It is a center that displays current scientific phenomenon within a sort of amusement park, designed to form a gigantic cultural conglomeration aimed at projecting the image of advanced industrial society. Science centers such as La Villette and the science, industry and technology "cities" have made it possible for the general public to approach and understand the scientific-technological universe.

A third possibility, more to entertain than for expert studies, can be found in the pavillions dedicated to science and the future in large recreational complexes. The Disney Corporation, for example, designed Epcot in Florida, a center that has become a focal point for all those enthusiasts who wish to experience a particular vision of what the future holds for the generations to follow. In France, the ambitious Futuroscope project in Poitiers is another contribution along these lines. More a spectacle than for intellectual reflection, its expressive potential converts it, nevertheless, as does Epcot's, into a first-class cultural instrument.

The first of the above models, the science museum, is thus characterized by the fact that it organizes, usually under one roof, a series of largely original technological and scientific artifacts. These artifacts generally date from the scientific-technological revolution—and are those to which can be attributed much of the rapid transformation of Western society. This preservation of the prototypes of many technological innovations that later became routine parts of daily life, along with models and reproductions, are viewed as didactic examples of the way in which those advancements influenced the society in which they arose. Older scientific materials are complementary to the collections and are often used as contrasting examples of the reality of other eras and ideas.

The second of these approaches, the science center, in contrast to the first, is practically devoid of any type of historical material, even though it can be found periodically in specific exhibitions. The idea or philosophy on which it was founded was that of presenting the scientific laws and effects that undergird and control the world today and that form the basis of modern technology to the lay person in accessible ways. Often utilizing ingenious, interactive computer-based machines designed to demonstrate the spectacular aspects of science with experiments chosen for their didactic value, the idea was to promote a dialogue between the visitor and the scientific/technological phenomena with which he or she was presented. At present this second philosophy is a very attractive one for a culture that reflects more on the future than on the past. Combined with the difficulties that mounting good historical exhibitions present, this makes for a very popular following for the interactive science center model. The extent to which this type of center has proliferated is highly indicative of a widespread sociological phenomenon. However, the way in which it has been employed as a complement to the traditional educational system and as a way of balancing the cultural differences mentioned at the beginning of this essay is a very controversial issue among museum specialists.[7]

INFORMATION, PROPAGANDA, OR SCIENTIFIC EDUCATION?

Both the science museums and centers are aimed at two major groups, the more prominent of which is the nonuniversity student sector, who view the science center as a complement to school studies, something that reinforces the scientific-technological knowledge that they have learned in the classroom. The other major group of visitors are those who, having completed their general education, use the center as a source of information or entertainment.

In the case of the first group, science museums have represented, up until now, an extension of the classroom. However, recent studies have confirmed that the extent of interaction between student and museum is much more complex than previously thought.[8] A science center is a complex environment in which the objectives, experiences, and situations are not the only things involved. The building itself—its characteristics, the atmosphere that it creates, the reaction to it by other visitors, and the interaction between what the students experience there and what they have been learning in their classrooms—captures 50 percent of the interest of the occasional visitor (on a guided tour) and an even higher percentage in the case of the more specialized visitor.[9] If a science museum tries to convert itself into a mere extension of the classroom and of the subjects taught there, the majority of the available information is lost or not considered applicable, in which case the institution has become yet another traditional school.

According to recent studies, to be most effective the teaching done in these centers should be of a very different nature. This type of installation, in order to carry out the function for which it was created, must be an open informational and experimental center in which resources can be freely used for ideas that have no necessary priority. The next step for these centers will be a focus on subjects of advanced research, which utilize specific mathematical and, especially experimental, techniques.[10] Beyond certain levels, the museum cannot substitute for academic study, but it can orient and prepare students for broad contextualized understanding of such complex material.

From an educational standpoint, the museum, science center, or science and technology "city" currently acts as an option to the traditional education system.[11] Students and teachers should view it as a new direction for education based principally on the teachings of science and technology. But these institutions are also the center of attention for new didactic approaches in science, in which the student interactively participates from the very beginning. This new method is not only useful for students, it also opens new possibilities for teacher education. The science and technology center is, or should be, an important instrument in teacher education and serves as a center for didactic experimentation to help prepare schools for the twenty-first century.

Although an extension to formal education is the crucial element for the first group of users of these installations, for those who have abandoned a traditional schooling in the area of science, either because of interest in other academic areas or because of other professional practices that do not involve a direct form of study, these institutions become nearly the only form of contact with the scientific-technological universe in which society is immersed. For this second set of citizens who comprise the large majority of the population and who should be making democratic decisions about the characteristics of the society in which they wish to live, about the risks they wish to take, what they are willing to tolerate, and, in the end, where science and technology should take future society, there are three main functions that the science museum should carry out: a source of information, a means of mass communication, and a prod to reflection. One more function could also be added this list, that of serving as a center for entertainment.

The world of science has always been an elitist and esoteric world. Until the end of the eighteenth century, science was able only to aspire to knowing the world. Except on very few occasions, the power of science (in contrast to technology) to transform the world was only a dream. This situation has changed radically in the past hundred years, and a dream has become a reality. Today we are dealing with the risk of nuclear and chemical warfare, the destruction of the ozone layer and of the forests, the eradication of biblical plagues and the discovery of new worlds. These risks entail important decisions that must be taken into account by the citizens of all countries in the world. To do so, today's citizens should have proper access to information that allows them to understand, even without expertise, the mechanisms that move the world, as well as the smaller instruments that make day-to-day life easier. This is undoubtedly the most difficult challenge that science centers face. While it is quite easy to mobilize the educational sector in this sense, for current educational trends favor such mobilization, it is much harder for science centers to reach the nonscientifically trained general public. The science center, if it really intends to be a popular route to understanding the scientific-technological world in which we live, must be provided with sufficient resources in order to attract those citizens who are not naturally inclined to reflect on this aspect of the world.

SCIENCE CENTERS AND TECHNOLOGY AND THEIR AREAS OF INFLUENCE

Today's Western societies are indeed technological societies, and many have also become societies of leisure. This scientific and technological world creates a great fascination for understanding in the average citizen, but it is a fascination often frustrated by the very complexity of the problems encountered and by the methods employed for the solution. The majority of these technically uninformed citizens are seldom interested in the details of any particular development, but they are interested in the general context, in spectacular achievements, and in important consequences. These concerns should thus be the object of special attention when designing displays for science centers.[12]

Technically based education has advanced so far today that it can readily depict the spectacular nature of scientific discoveries and technological advances, while allowing visitors to observe and experiment with previously "secret" aspects of the world and daily life.[13] A science center, in line with its objectives, should thus employ all of the technological resources at its disposal to make its displays as spectacular and exciting as possible. New generation planetariums, advanced projection systems, interactive information networks, powerful instruments for observation and exhibition, and at the same time, realistic settings and appropriate theatrical recreations should all be utilized to present valuable data, attractive information, and didactic as well as amusing shows to fill the hours of leisure and answer even the most detailed questions.

Emphasizing the more spectacular aspects of the scientific-technological world and creating "scientific shows" will attract even those citizens who are not especially interested in learning in great depth about the universe. A visit to a science center should not and, in fact, cannot be a substitute for written texts. At the same time, science centers should not present written types of explanations. Experience has shown that in the case of voluntary visits, time spent reading written explanations decreases rapidly over the period of the visit and ends up being quite partial.[14] Science centers should be, first and foremost, institutions that promote an interest in the area of science and technology through didactic instruments of powerful but, at the same time, entertaining dimensions.

The science and technology center, despite whatever philosophy might prevail, and despite its good intentions, will always have difficulty in reaching the average citizen. The greatest obstacle to reaching large numbers of people is centralized location; large-scale dimensions and high costs of construction and maintenance are incompatible with widespread proliferation. Although this is a general problem that also arises for centers of medicine, universities, and cultural and leisure activities, it is often mitigated by the mobility of the population. Thus, the science and technology center must lure the average citizen to the center; this will require new initiatives.

There are two measures that will likely produce the greatest effects. The first of these would involve the decentralization of the installation to the greatest possible extent, bringing it as near as possible to citizen residences. Thus, one might think along the lines of creating peripheral units around a major city, especially in those areas with past and/or present scientific or industrial traditions. A satellite system and a network of associated museums would complete the picture.

The second measure would entail creating an itinerant center, which would permit the transfer of displays to population centers that can provide the minimum appropriate setting. The present educational system shows great possibilities in this area, which could be complemented by creative ideas such as mobile scientific video clubs and self-contained itinerant exhibitions, for which the entertainment world has already created models, but which unfortunately are not fully appreciated nor utilized to the extent they could be. The idea is to bring the science center closer to the average citizen, to provoke his curiosity, and, ultimately, to close the mental and physical gap that separates the average person from the environment in which he or she lives.

In summary, if technology in our society is increasingly a one-way street, and if increasing specialization threatens to put further distance between the average citizen and an understanding of the basis of this technical work, then a special effort must be made to find the means for salvaging the imbalance between utilization and comprehension and to prepare future generations in this respect. The creating of specialized science and technology education institutions through which the scientific-technological world both can be demonstrated and understood in its full reality is one such readily available means. Although the details of the problem must be discussed before we will be able to apply this approach to the real world of people, it does offer a new idea for responding to new technologies in a rapidly evolving world.

NOTES

1. The interest in science museums has already resulted in a considerable bibliography on their conception, characteristics, and utilization as educational instruments. S. A. Bedini's "The Evolution of Science Museums," *Technology and Culture* 6 (1965): 1–29, is still useful in this context.

2. E. Bonnefous, *Le Conservatoire national des arts et métiers. Son histoire, son musée* (Paris: CNAM, 1987).

3. A. Rumeu de Armas, *Oriegen y fundación del Museo del Prado* (Madrid: Instituto de España, 1980).

4. J. Teichmann has studied the didactic origins of the Deutsches Museum and the evolution of the philosophy of its activities. The Deutsches Museum and the courses given in their installations have become an important point of reference in scientific museum studies. See "Deutches Museum, München—Science, Technology and History as an Educational Challenge," *European Journal of Science Education* 3 (1981): 473–78.

5. See D. Chisman, comp., *UNESCO Sourcebook for Out-of-School Science and Technology Education* (Paris: UNESCO, 1986), for a good summary of recent essays on science and technology education.

6. F. Oppenheimer, "The Exploratorium: A Playful Museum Combines Perception Arts and Science Education," *American Journal of Physics* 40 (1972): 978–84; and Oppenheimer and K. C. Cole, "The Exploratorium: A Participatory Museum," *Prospectus* 4 (1974): 4–10, are still useful introductions to this interesting and successful experience.

7. For further information on the educational role of museums, see V. J. Danilov, "Science Museums as Educational Vehicles," *Science Teacher* 40 (1973): 26–27; G. McGlaherty and M. N. Hartmann, "The Museum as a Teaching Resource: An Inquiry Approach," *Science and Children* 11 (1973): 11–13; and R. Bud, "The Myth and the Machine: Seeing Science through Museum Eyes," *Sociological Review Monograph* 35 (1988): 134–59.

8. J. J. Koran, Jr., S. J. Longino, and L. D. Shafer, "A Framework for Conceptualizing Research in Natural History Museums and Science Centers," *Journal of Research in Science Teaching* 20 (1983): 325–39; and A. M. Lucas, P. McManus, and G. Thomas, "Investigating Learning from Informal Sources: Listening to Conversations and Observing Play in Science Museums," *European Journal of Science Education* 8 (1986): 341–52.

9. J. H. Falk, J. J. Koran, Jr., and L. D. Dierking, "The Things of Science: Assessing the Learning Potential of Science Museums," *Science Education* 70 (1986): 503–8.

10. We have studied this problem in the teaching of astronomy. See A. E. Ten and M. A. Monros, "Historia y enseñanza de las Astronomía I. Los primitivos instrumentos," *Enseñanza de las Ciencias* 2 (1984): 49–56; "Historia y enseñanza

de la Astronomía II. La posición de los cuerpos celestes," *Enseñanza de las Ciencias* 3 (1985): 47–56; and "La Astronomía y su enseñanza," *Comunidad escolar* 3, no. 40 (1985): 13.

11. M. C. LaFollette et al., "Science and Technology Museums as Policy Tools—An Overview of the Issues," *Science, Technology & Human Values* 8 (Summer 1983): 41–46. Also helpful is U.S. Congress, Office of Technology Assessment, *Elementary and Secondary Education for Science and Engineering: A Technical Memorandum* (Washington, D.C.: Office of Technology Assessment, 1988).

12. V. W. Stern et al., *Out-of-School Programs in Science* (Washington, D.C.: American Association for the Advancement of Science, 1981).

13. Some recent examples from among the wide range of information now available can be found in D. A. Ucko's "An Exhibition on Every Day Chemistry: Communicating Chemistry to the Public," *Journal of Chemical Education* 63 (1986): 1081–86 and J. R. Bird's "The Potential of MEV Ion-beam Techniques in Museum Science," *Interactions with Materials and Atoms* 14 (1986): 156–61.

14. Falk, Koran, and Dierking, "The Things of Science," 504.